Urban Fields

A Geometry of Movement
for Regional Science

S Angel
GM Hyman

Urban Fields

A Geometry of Movement for Regional Science

S Angel
G M Hyman

p Pion Limited, 207 Brondesbury Park, London NW2 5JN

ISBN 0 85086 052 0

Printed in Great Britain by J. W. Arrowsmith, Bristol

75 014309

Preface

The way in which we move about in our cities relies on our understanding and experience of distances and times. When road conditions are about the same everywhere, distances are roughly proportional to travel times. The most direct routes are then simple straight lines, and we can usefully perceive urban space as a smooth geometrical plane. With the rapid growth of metropolitan areas, this simple geometrical framework becomes less and less useful. The uneven growth of transportation systems in modern cities has resulted in great variations in the ease of movement in different parts of the city. Traffic can now move at unparalleled speed in some locations, and at a tortuous crawl in others.

This important development naturally requires a new perception of space, one in which direct routes are no longer straight lines, and distance is no longer measured with a ruler on a city map. Indeed, such a perception is now commonplace among urban populations.

The aim of our book is to study this new perception of urban space with the more precise tools of scientific investigation. By making use of mathematics, differential geometry in particular, it becomes possible to create a framework for the analysis of a number of important urban phenomena in their natural setting—their spatial, geographical framework. The lack of a more satisfactory geometrical framework for the study of urban and regional space has prompted many scholars to abandon the geometrical framework completely. In recent years the most important contributions in land-use planning and transportation studies have been based on numerical approaches, greatly facilitated by the introduction of the electronic computer. These studies have all rejected the early simplistic plane geometry; it has been replaced by a spatial framework involving the division of geographical space into a set of zones, and by the representation of the transportation system by a network. Such a system requires no geometrical representation.

The numerical approach enables us to describe and analyze a greater structural variety of cities and transportation systems than had been previously accommodated in the earlier geometrical representation of space. However, the perceptions of general patterns and principles of the organization of space in urban regions—the most important insights into earlier developments of regional science, urban economics, and regional geography—have become difficult to find in the cumbersome detail of the numerical framework.

The increasing complexity of urban systems requires a more complex geometry to describe it. This book investigates the different requirements of such a geometry and demonstrates, with several important examples, how it can adequately describe a number of important properties and processes in urban areas. With such a geometry it becomes possible to integrate many findings concerning the spatial structure of cities, and the spatial interaction within them. It thus enables us to overcome the impasse which has lead to the earlier abandonment of the geometrical approach, and gives the geometrical treatment of geographical phenomena a new lease on life. On the other hand, it makes it possible to study general patterns of space in urban systems, thus increasing the usefulness of numerical studies. In essence such a geometry is crucial to the improvement of our understanding of the spatial aspects of the variety of urban phenomena that shape our lives.

The book is divided into two parts. In Part 1 we present our own work on the subject, beginning with a systematic exposition of a geometry of movement for urban areas, and continuing with the application of this geometry to a number of problems in regional science and related disciplines. In Part 2 we present a number of key articles which closely relate to our work, and which serve to give the reader a comprehensive picture of the subject, its importance, and its possibilities. All of the articles presented in Part 2 are closely related to the various chapters of Part 1, and are referred to a number of times in the text. Having these papers close at hand is of considerable help, since copies are often hard to find. The first two articles, by Professor Warntz, set out in the most general and visionary terms the breadth of the modern geometrical perception of urban and regional space. Both are potent with ideas, many of which are exciting and challenging to an interested researcher. At the same time, they provide useful basic examples of the applications of these ideas to concrete geographical problems like finding minimum paths and isochrones on a world map, like finding the shortest time path of transatlantic flights (given the wind patterns over the ocean), and like finding the cheapest land acquisition map for road building. These examples serve to convey in very clear terms the direction of thinking and research on the subject, and have helped us greatly in forming our own ideas. In short, Professor Warntz's insights permeate our book.

The article by Professor Tobler is an excellent introduction to the subject of map projections, and particularly to the use of such projections for the description of human activity in geographical space. Professor Tobler gives a number of interesting examples of such projections—maps designed to display retail sales, population, or migration instead of simple geographical areas. The article by Professor Wardrop applies the idea of map projections to transportation and accessibility in urban areas. These two articles, together with chapter 3 provide the reader with a broad perspective of the role and use of map transformations in urban and regional studies.

Finally, the article by Professor Wilson provides a broad and systematic presentation of a large class of urban interaction models, models of movement and exchange in urban space. This article is an excellent example of the advances obtained by the introduction of the numerical approach into descriptive and predictive urban studies. The paper forms the basis for the construction of the geometrical model of spatial interaction presented in chapters 4 and 5. Again, reading these in parallel makes it possible for the reader to follow our line of development, and to see for himself where he may go from here.

Thus the book provides a framework which integrates the conceptual and empirical advances in land-use and transportation studies with the substantive theory developed in the academic disciplines.

Most of this book is of direct interest to teachers and students in the field of regional science. The three chapters following the introduction are of especial interest to quantitative geographers and transportation students, the initial chapter being adaptable to a series of special projects involving the detailed analysis of transport and land-use characteristics of urban areas. The mathematical appendix is of particular relevance to these studies. It provides both a source of analytical techniques which supplements those normally available to geographers, and a language for assessing the results of this analysis in the context of current theory. They are also of practical significance in the development of forecasts of the impact of possible changes that may occur in the urban system. The final chapter will be of particular interest to environmental researchers, urban economists, and students in related subjects who are concerned with the economic analysis of location within the urban area.

S. Angel
Asian Institute of Technology, Bangkok
G. M. Hyman
Centre for Environmental Studies, London

Acknowledgements

We must acknowledge our appreciation of all those people who have helped us with this book. The director of the Centre for Environmental Studies, Professor David Donnison, and our colleagues at the Centre provided us with a comfortable and efficient environment during the two years of research. In particular we have been greatly assisted by criticisms and support from Professor Alan Wilson, who also encouraged us to publish some of the early results. We have also benefited from comments from Dr Martyn Cordey-Hayes. Professor William Alonso, Professor Michael Teitz, and Professor Roland Artle in the Department of City and Regional Planning, University of California, Berkeley provided valuable guidance and commented on earlier versions of this work. Professor Waldo Tobler gave us important suggestions. Professor Walter Isard and Professor David Boyce displayed genuine interest in our work and made it possible to present some of the early results in conferences of the Regional Science Association. The data for the study were obtained from SELNEC (South East Lancashire–North East Cheshire) Transportation Study, through the Mathematical Advisory Unit of the Ministry of Transport in London. Anthony Hawkins and Michael Hammerstone gave us valuable assistance in obtaining and interpreting the data.

Much of the material published in this book is reworked from papers published in various journals. We would like to acknowledge the editors and publishers of several journals in this respect. Some of the results of chapter 2 and chapter 4 were published in *Environment and Planning* (Angel and Hyman, 1970; 1972a), and some of the material of chapter 2 and chapter 5 was published in *Papers of the Regional Science Association* (Angel and Hyman, 1971a; 1972b). The material of chapter 3 was published in *Geographical Analysis* (Angel and Hyman, 1972c).

Finally we should like to record their grateful thanks for permission to reproduce material appearing in this book, as follows:

To Professor William Warntz for permission to republish his article "A note on surfaces and paths and applications to geographical problems"—our pages 121–134.

To the Regional Science Association for permission to reproduce "Global science and the tyranny of space", by W. Warntz, *Papers of the Regional Association*, **19**, 7–19, 1967—our pages 108–120.

To the Minister for Transport of the German Federal Republic, for permission to reproduce "Minimum-cost paths in urban areas", by J. G. Wardrop, in *Beiträge zur Theorie des Verkehrsflusses*, Eds W. Leutzbach, P. Baron, *Strassenbau und Strassenverkehrstechnik*, **86**, 184–190, 1969—our pages 155–161.

To the American Geographical Society for permission to reproduce "Geographic area and map projection", by W. R. Tobler, *Geographical Review*, **53**, 59–78, 1963—our pages 135–154.

To Pergamon Press Ltd for permission to reproduce "A statistical theory of spatial distribution models", by A. G. Wilson, *Transportation Research*, 1, 253–269, 1967—our pages 162–178; also a figure from *Von Thünen's Isolated State*, 1966, Ed. P. Hall—our figure 3.2.

To Edward Arnold (Publishers) Ltd, for permission to reproduce a figure from *Locational Analysis in Human Geography* by P. Haggett, 1965—our figure 2.1;

a figure from *Network Analysis in Geography* by P. Haggett and R. J. Chorley, 1969—our figure 2.9;

a figure from *Explanation in Geography* by D. Harvey, 1969—our figure 3.5.

To MIT Press, Inc., for permission to reproduce two figures from *Location and Space Economy* by W. Isard, 1956—our figures 3.3 and 3.4.

We should also like to thank Professor Walter Isard for allowing the latter of these two figures to be used as a basis for the jacket design.

Contents

1

Introduction

Traffic congestion is a contagious phenomenon. If one location is heavily
congested, then another location nearby will also become heavily congested.
This phenomenon has been characterized in the traffic-flow literature as
Wardrop's Principle of Equal Time (Wardrop, 1952).

> "As one route becomes congested, vehicles switch over to another,
> faster route. As more drivers change to the other route, speeds on that
> route decrease. Equilibrium is reached when travel on either route costs
> the same. The local effect of this phenomenon is to spread traffic
> among the roads in a given area in such a manner so as to eliminate
> differences in road speeds in the area".

The approach which we shall adopt in this work assumes that the time or
cost of movement is determined by one's location. This means then that,
broadly speaking, a traveller cannot determine his (or her) speed of travel,
accelerate, or decelerate, even when driving his own vehicle. The conditions
for travel are fixed externally by the general conditions of a particular
location—the roads, the traffic density, the speed limit, the weather, and
so on. We associate with any location within the region under study some
average speed or average cost of travel, and we treat velocities or transport
costs as functions defined over a map of the region. Given this simple
assumption we can represent travel as taking place along minimum paths,
that is to say paths of least time or paths of minimum cost, and follow
the suggestion of William Warntz:

> "The general analytical approach to the study of such paths as these of
> economic and social significance is through differential geometry and
> the calculus of variations, and the principal concepts employed lie
> within a generalized theory of refraction, a general spatial *lex parsimoniae*.
> Thus, for example, to establish the minimum land-acquisition-cost paths
> between two points within a region, we assign to land values, when
> integrated around a given point, a role isomorphic with that of an index
> of refraction" (Warntz, 1967, p.8—see page 108 of this book).

As will be seen later, these minimum paths will not usually be straight
lines, that is, paths of least distance on the Euclidean plane. Two points
which are on opposite sides of a congested city centre may be ten minutes
apart when one travels straight across the centre, but seven minutes apart
with the optimal detour about the centre. The best route will be along
the path which entails the least expenditure. When the cost of transport
is distributed smoothly on the Euclidean plane, the optimal path will be a
smooth curve on the plane. This path is an abstraction of the form that
the path traces on the road network, where the kinks and turns have been
smoothed out and only the general direction of travel can be discerned.

The geometry of transport which we shall develop here is based on this one simplifying assumption concerning the distribution of transport costs in space. The cost of travelling a unit distance is assumed to be a function of location and therefore to be independent of the direction of travel. We further characterize the cost of travel as related to location in a wider sense. The cost per unit distance of travelling at any location is assumed not to vary appreciably from the cost of travelling a unit distance at a location very close by. The distribution of costs over geographical space is thus assumed to be smooth.

The distribution of velocities of travel on the Euclidean plane will be referred to as a *velocity field*. We shall make no distinction between speeds and velocities. We subsequently use the term 'velocity' for ease of expression, although generally it will not imply a specified direction of movement.

In the following chapter we shall create the geometrical framework for studying movement in urban and regional space. We shall describe the necessary assumptions which are needed to construct velocity fields, and various methods of constructing and visualizing these fields. Urban velocity fields, portraying the traffic situation in modern congested urban areas will be constructed for three British cities: Manchester, Glasgow, and London. The special distinguishing features of urban velocity fields will be discussed in detail—they all display fairly systematic variations in average velocities as the distance from the centre of the city increases.

Having obtained the general characteristics of velocity fields, we shall develop procedures for obtaining the shape and form of paths on these fields. Following traditional practices, we shall focus our attention on minimum paths—paths of least resistance, least time, or least cost. Given these paths, it is possible to calculate travel time or travel cost between any pair of locations. More generally, it is possible to obtain the shape of isochrones and isocost contours—the loci of all points which are an equal time or an equal cost away from a given point. These will be illustrated by examples in the text, and by a mathematical analysis at the end of the chapter.

The theoretical analysis of the form of minimum paths and isochrones, and the calculation of travel time, is largely adapted from modern mathematical physics. Its application to quantitative measurements of fields and paths in urban regions is novel. In order to facilitate the quantitative analysis, we have simplified the measurements of fields. We have limited the analysis to radially symmetric fields which are constructed by plotting the average velocity at varying distances from the city centre. This representation allows for a considerable simplification of the mathematical analysis, and has obvious advantages for the display of information. To test the validity of the analysis, we have compared the measurements obtained for travel times between pairs of points in the field with measurements of travel time on the road network. The comparison is highly satisfactory, despite the simplifying assumptions of smoothness and radial symmetry.

Many previous studies that have attempted a geometrical explanation of the spatial structure of cities and regions have had to adopt simplifying assumptions in order to derive their results. Those which are frequently made are the assumption of the Euclidean plane, the assumption of uniform densities, and the assumption of uniform transport facility. Classical works on regional patterns, such as the works of von Thünen (1966, first published 1875), Christaller (1966, first published 1933), and Lösch (1967, first published 1940) make all three assumptions. Other authors make various combinations of these assumptions.

The representation of transport costs by a velocity field allows us to examine critically the idea that regional geographical space, with its nonuniform densities and its nonuniform distribution of transport costs, can be transformed into some abstract space where some or all of these assumptions hold. The existence of such transformations is usually proposed as a means of relating the patterns predicted by these theories to observed geographical patterns. In chapter 3 we shall study four conjectures concerning the existence of such transformations of geographical space, each named after the author primarily associated with it. First, we shall review the conjecture that there exist transformations which map a Euclidean plane with a non-uniform distribution of densities to a Euclidean plane with a uniform distribution of densities (Tobler's conjecture). Then we shall show that transformations exist which map the Euclidean plane, with a nonuniform distribution of transport costs, given as a field, into a curved surface on which these costs are evenly distributed (Warntz's conjecture). We shall study the construction and properties of such surfaces and discuss several examples. A realistic example of a curved time surface will then be derived for the Greater Manchester velocity field. Movement on such surfaces takes place along geodesic paths (shortest distance paths). All characteristics of minimum paths in the field can be obtained by studying the geodesic paths on the surface associated with it.

The transformation of a velocity field into a curved surface, to be referred to as a *time surface*, allows us to study two further conjectures. We shall show that no transformation exists which allows us to retain both the assumption of the Euclidean plane and the assumption of uniform transport facility (Wardrop's conjecture). A surface obtained from a realistic field will in general be curved. We shall also show that both the assumption of uniform densities and the assumption of uniform transport facility cannot be retained simultaneously, even on a curved surface (Bunge's conjecture).

We shall thus conclude that urban and regional theories which require combinations of assumptions for which transformations do not exist, such as the theories of Lösch (1967) and Christaller (1966), can only be adequately tested if their spatial assumptions are relaxed.

In chapter 4 we shall introduce a geometrical theory of spatial interaction. A model of spatial interaction is derived which is calibrated for the Greater Manchester urban area, 1965, using data from the SELNEC (South East Lancashire–North East Cheshire) Transportation Study. The model is applied to reproduce the distribution of trips—allocating commuter trips from any area to any other area in an urban region in accordance with the derived trip-density function. The required inputs for the model are a density function for residences of commuters, a density function for workplaces of commuters, and a velocity field for a given urban region. The velocity field is used to calculate travel time between locations. The predicted travel times are shown to be remarkably similar to those obtained by the SELNEC Transportation Study—the SELNEC values being derived from a network representation of the transport system. The trip-density function is used to obtain several surfaces defined over the urban region. In particular we obtain the surface characterizing the accessibility of workers to jobs, the surface characterizing the accessibility of workplaces to residences, and a surface characterizing the spatial distribution of traffic.

This model allows us to investigate another branch of geometrical theories of urban spatial structure, the economic theories of urban rent. The works of Alonso (1965), Mills (1967, 1969), Muth (1961, 1969), and Wingo (1961) will be examined in this context. These authors make assumptions, concerning expenditures on transport, based on the premise that all destinations of commuter trips are concentrated in the Central Business District (CBD). In chapter 5 we are able to show that these assumptions lead to contradictions and that the premise that all destinations are concentrated in the CBD needs to be relaxed in order to obtain a realistic representation of expenditures on transport.

Fields: a geometry of movement

2.1 The construction and properties of urban fields

This chapter is concerned with creating the geometrical framework for studying movement in urban and regional space. This framework takes geographical space to be a plane on which travel costs or speeds are distributed as continuous functions—these functions are referred to as *fields*. We shall first describe the important properties of fields and give several empirical examples of urban fields.

We shall then describe a graphical method and a computer algorithm for the calculation of travel time between locations in a given field. These methods will be used to obtain estimates for travel times on the velocity field constructed for Greater Manchester, 1965. When these estimates are compared with measurements of travel times on the road network, they are shown to be satisfactory despite the simplifying assumptions of smoothness and radial symmetry.

Using the methods described in this chapter, the reader should be able to construct fields for any given city or region and to obtain theoretical approximations of the form of paths between any locations in these fields. It should also be possible for the reader to construct good time estimates or cost estimates. Such constructions will be found useful in giving a simplified global picture of the transportation behaviour of the entire city or region, unencumbered by the detailed descriptions of road networks which are usually encountered.

A smooth function which assigns velocities to locations in the geographical plane will be referred to as a *velocity field*. A given velocity field will generally relate to a specified mode of travel. When we are concerned with the costs of travel per unit distance, we shall speak of a *cost field*. One way of representing travel costs is by the distribution of travel time per unit distance. Instead of referring to the cost per mile at any location, we could refer to the travel time per mile. Thus, if the velocity of travel is sixty miles per hour, the cost per unit distance would be one minute per mile. When we are concerned with the distribution of time per unit distance, we shall speak of an *inverse velocity field*.

We shall assume that velocity fields have certain definite properties as follows: (1) travel can take place anywhere and in all directions, (2) there is a unique velocity associated with each location, independent of the direction of travel, and (3) the velocity of travel varies smoothly between nearby locations. The first property makes the analysis more suited to modes of transport that are not restricted to a small number of paths. Thus urban road systems and private cars appear more amenable to study than rail transit. Urban fields can therefore be represented as smooth functions over the geographical plane.

The general form for an urban field may be visualised as a hilly topography. In a velocity field the hills and ridges would represent regions of high velocity, and the valleys congested regions where the velocity is low. Such hypothetical urban field is illustrated in figure 2.1[1].

The bulk of the analysis given in this chapter—the study of the characteristics of travel on minimum paths—could be applied to the forms of urban fields depicted in figure 2.1. There are several alternative methods for describing the characteristics of a minimum path in such a field. One is analogous to the method developed by Huygens (1912, first published in 1678) to determine the wavefronts of light. Another is based on a soap bubble model, where the urban field is represented as a rigid surface lying above the urban plane. These methods will be discussed in greater detail below. In chapter 3 a different method is described, where minimum-time paths on the urban plane are represented as shortest paths on a curved surface.

These methods enable the whole system of minimum paths to be described in general qualitative terms. However, when we wish to make detailed calculations of travel time there are practical limits to consider. We were therefore led to make one further major simplification whenever particular quantitative results were required in this study. This simplification is the *assumption of radial symmetry*. When this assumption is made we regard the velocity of travel as a function of distance from the city centre.

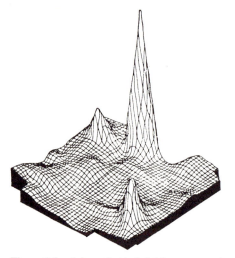

Figure 2.1. A hypothetical field represented as a smoothed statistical surface (drawing from Haggett, 1965; after Jenks, 1963).

[1] There are several methods for representing discrete field data as a smoothed statistical surface. The problem has been discussed by Tobler (1963) who cites several statistical references on the subject. A recent review of programs used for the construction of contours from discrete data appears in Rhind (1971).

Thus a unique velocity is assigned to the perimeter of each circle, in the urban plane, whose centre coincides with the centre of the city. In those cases where radial symmetry is not required, as in the discussion of transformations in chapter 3, or in the development of the continuous model of spatial interaction in chapter 4, the general form for urban fields, mentioned previously, can be assumed.

By focusing on radially symmetrical cities, we are following the tradition of many urban analysts. We have to distinguish, however, between the assumption of radial symmetry as discussed here and other models of radially symmetrical cities discussed in the literature. For instance Isard and Liossatos (1972a) cite levels of air, water and solid-waste pollution, the probability of crisis conditions, traffic congestion, land rents, shipping costs, access to high-quality goods and services, and the cost of commuting as depending only on distance from the city centre. Park and Burgess (1925) have developed a concentric ring model of urban growth, where the city is divided into five concentric zones. Alonso (1965) has applied von Thünen's concentric ring theory of agricultural rents in an urban context, the city being divided into concentric rings of infinitesimal width which are the equilibrium amounts of land consumed by households and firms. Concentric ring models have also been applied by other urban economists, notably Mills (1967, 1969), Muth (1961, 1969), and Wingo (1961).

In these models, however, travel is always assumed to take place on radial routes to and from the city centre. The two-dimensional plane of the city does not, therefore, play an important role. The analysis and the display of information are limited to a line, any straight line proceeding away from the centre. Areas and areal distributions are thus reduced to lengths and cross sections. Movement reduces to straight line movement along radial paths.

The assumption of radial symmetry also follows a tradition in the transport literature, where an established body of work is devoted to circular cities. Analysts try to measure the different characteristics of transportation system by making various assumptions about the arrangements of roads. Mean journey lengths on various road systems (Holroyd, 1966), the traffic flow on radial and circumferential routes (Lam and Newell, 1967), and the proportion of space occupied by roads (Owens, 1968) have been discussed in this framework. These studies often allow for travel in all directions without restricting it to radial paths, thus resembling the approach to be developed here. While following a long-standing tradition, the assumption of radial symmetry does also appear to produce reasonable predictions of travel time.

Let us now investigate the form of velocity fields, mainly by way of example. The simplest possible velocity field is obviously the *constant-velocity* field, where the speed of travel is everywhere the same. This velocity field can be represented by the Euclidean plane. Many studies in regional geography implicitly use the constant-velocity field as their representation of the transportation system in the region under study. In computing, say, coal transport prices for a region as large as the United States, it is quite safe to assume that transported coal travels in approximately straight lines at a fairly constant average speed. The same may be true for aircraft at large distances from congested airports, or for boats. In modern urban areas, though, it has long been recognized that the constant velocity field can be quite misleading. Straight lines are bad representations of optimal routes, and distances usually fail to give good estimates of travel times. This is particularly true of motor traffic, which is the focus of this study. Study of movement in existing urban areas indicates that realistic fields possess certain common properties. The velocity of travel is found to be at a minimum near the city centre—the location of maximum congestion. This velocity is found to increase with distance from the city centre and, for a given mode, to level off as distance from the centre increases. The above observations relate to empirical measurement of velocity fields in three British cities, Manchester, Glasgow, and London.

A cross section of the velocity field for Greater Manchester, 1965, is shown in figure 2.2. The velocity field was obtained by sampling vehicle speeds on the existing road network, using peak-hour travel data from the SELNEC study (SELNEC, 1968). A random sample of velocities was drawn at various distances from the city centre, and the results at each distance were averaged. These averages are represented by the dots in figure 2.2. A smooth function, $V(r) = a - b\exp(-cr)$, was fitted by requiring it to pass through three chosen points, where r is measured in miles and V in miles per hour.

Significant deviations from the curve, particularly those at 3 and 6 miles from the centre, appear to be due to congestion in the rings of dense

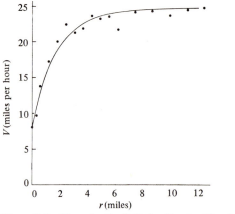

Figure 2.2. The velocity field for Greater Manchester, 1965.

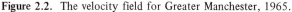

settlement outside Manchester: Urmston–Sale–Stockport and Bolton–Bury–Rochdale–Oldham.

A cross section of the velocity field for Glasgow, 1961, is presented in figure 2.3. This was obtained by the same procedure as described above, using data from a Glasgow highway study (Glasgow Corporation, 1965).

Although the measurements for Glasgow were not available for distances beyond 5 miles from the city centre, velocities seem to be levelling off beyond this range, and the distribution of velocities is similar to that of Manchester. A negative exponential function could also be fitted for the Glasgow data, yielding the velocity field $V(r) = 24 \cdot 8 - 18 \cdot 0 \exp(-0 \cdot 75r)$. While speeds on the periphery are almost identical with those in Manchester, central speeds are lower, but increase faster as distance from the centre increases.

Similar measurements were taken in London[2]. However, while in each of the previous cases morning peak-hour measurements were obtained, the measurements for London are for the daytime off-peak period. The results for London (figure 2.4) are considerably different from those of the previous two examples. The average central velocity (14 mph) appears to be much higher than that for Manchester (8 mph) or Glasgow (6·8 mph). Velocities also increase at a much slower rate, but reach a maximum of over 30 mph. However, the function used previously does not fit the

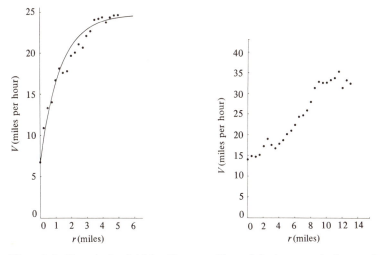

Figure 2.3. The velocity field for Glasgow, 1961.

Figure 2.4. Average velocity as a function of distance from the centre of London, 1969.

[2] The centre of London was taken to be Aldwych. Data for the measurements were obtained from the Planning and Transport Department, Greater London Council by Carole Pountney. The speeds are 'moving-car observer' speeds for 1969.

London results and no attempt was made to fit a function to London's velocity field. The Manchester velocity field will serve as a basic empirical field throughout this study.

These empirical measurements indicate in broad terms the general form of velocity fields in urban areas. A systematic comparison of the velocity fields for different regions provides us with a concise way of representing the relative performance of the vehicular transport system in these regions.

2.2 Movement along minimum paths

Velocity fields provide us with a framework for studying the movement of traffic. We assume that traffic moves along a *minimum-time path*, defined as a path between two points on the plane which can be traversed in the smallest time. The set of points on the plane which can just be reached in a given time from a fixed point is referred to as an *isochrone*. In a similar way we shall refer to *minimum-cost paths* and *isocost contours*. Where a minimum-time path and an isochrone cross each other their tangents are at right angles. The family of minimum-time paths and the family of isochrones for a given point are therefore orthogonal.

This can best be illustrated by example. In the constant-velocity field, $V(r) = V_0$, mentioned earlier, the family of minimum-time paths from a fixed point is the family of all straight lines emanating from that point. The corresponding family of isochrones will be the family of all the concentric circles about that point. The two families are clearly orthogonal to each other. The family of minimum-time paths and isochrones for the constant-velocity field is presented in figure 2.5 for a point east of the city centre.

In the case of the constant-velocity field the minimum-time path between two points coincides with the minimum-distance path. Minimum-distance

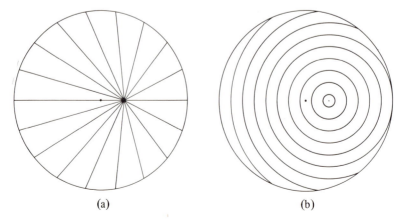

$$\text{(a)} \qquad\qquad\qquad\qquad\qquad \text{(b)}$$

Figure 2.5. (a) Minimum-time paths and (b) isochrones for the constant-velocity field, $V(r) = V_0$.

paths have long been an object of study of geometers and regional scientists. A common problem in network analysis is to determine a pattern of minimum-distance paths between a set of points in such a manner as to minimise the total length of the links (Haggett and Chorley, 1969). If the number of points to be connected is not too large the solution may be obtained by using a film of soap. The points are represented by pegs which are sandwiched between two parallel planes, one of which is transparent, and both planes are immersed in a soap solution. When removed, soap films are retained perpendicular to the planes which form the pattern of minimum overall link length. The junctions between films are at 120 degrees. Figure 2.6 illustrates the kind of pattern that results. The required pattern is obtained because the soap film assumes the minimum possible area. This property will enable us to generalise the method to construct minimum-time paths for a nonuniform velocity field. In this case, the planes will no longer be parallel to each other.

A similar principle provides the basis for an analogy between the determination of the least-cost route between two points and the path of a ray of light. This arises because light rays obey Fermat's principle of following the path of minimum time. Consequently Snell's law for the refraction of light can be applied to the problem of the determination of the route of minimum travel cost, so we can derive a *law of refraction of traffic*. The following derivation is due to Lösch.

Suppose that, in figure 2.7, a product is to be shipped from the point A to the point B. South of the coastline CD which is equally favourable everywhere for landing, the rail freight rate, f_b, is in effect; north of it the cheaper ocean freight rate, f_a, is in effect. The transport cost per unit

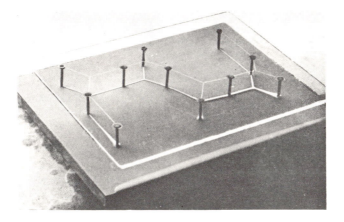

Figure 2.6. Soap film solution of link-length minimisation problem involving eleven points (based on Miehle, 1958).

for the distance AE will then be

$$F_a = f_a \, [a^2 + (c-x)^2]^{1/2} \; ;$$

and for EB,

$$F_b = f_b \, [b^2 + x^2]^{1/2} \; .$$

The total transport cost,

$$F(x) = F_a + F_b \; ,$$

is a minimum when

$$F'(x) = f_b \sin\beta - f_a \sin\alpha = 0 \; ,$$

where $F'(x)$ is the derivative of $F(x)$ with respect to x.
Hence

$$\frac{\sin\alpha}{\sin\beta} = \frac{f_b}{f_a} \; .$$

This determines the site of the harbour E. If the local velocity of light is substituted for the inverse of the freight rate in this formula, we obtain Snell's Law for the refraction of light. Thus the cheapest route from A to B will be bent at the coastline in the same way that a light ray changes direction when going from air to water: one follows a minimum-cost path through regions of differing freight rates, the other follows a minimum-time path through regions with differing local speeds of light.

In view of the analogy between minimum-time paths and light rays there is a corresponding analogy between isochrones and wavefronts of light. The correspondence arises because, just as the minimum-time paths are orthogonal to the isochrones, so the light rays are orthogonal to the wavefronts. We can therefore apply Huygens' construction for the wavefronts of light to obtain the form of isochrones. Figure 2.8, reproduced

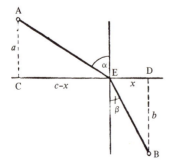

Figure 2.7. The law of refraction.

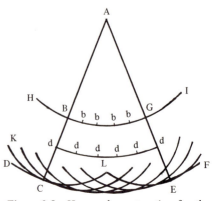

Figure 2.8. Huygens' construction for the wavefronts of light from a point source.

from Huygens' original text (but relettered), illustrates the construction of the spherical wavefronts arising from a point source in a medium of constant refractive index.

The principle employed in Huygen's construction is that each point on one wavefront, such as dd, can be regarded as a source for the construction of the wavefront occuring a unit of time later. So we construct a series of spherical arcs, such as KCL, from each point d on the old wavefront. The envelope DCF described by these arcs determines the form for the new spherical wavefront.

The case of a point source in a medium of constant refractive index is the simplest possible situation and corresponds to the case of the constant-velocity field discussed earlier. However, Huygens' construction can be directly applied to a medium of varying refractive index. Thus we shall be able to employ Huygens' construction to obtain the form of isochrones for the types of velocity field commonly encountered in urban areas.

An illustration of a family of minimum cost paths and the orthogonal family of isocost contours is provided by Warntz (1965). In this example, presented in figure 2.9, the measure of cost per unit distance is the cost of acquiring stretches of land for the construction of a highway. The isocost contours in figure 2.9 indicate the total cost of acquiring the necessary land between the contour and the centre, α, (Murfreesboro, Tennessee) along a minimum-cost path to any location in the United States. Six minimum cost paths are shown in broken lines.

A simple way of visualising the problem of finding a minimum-time path is to consider the inverse velocity field as a surface lying above a map of the city. A normal to the map at any point cuts the surface of the inverse velocity field at a corresponding point. As we move along a line

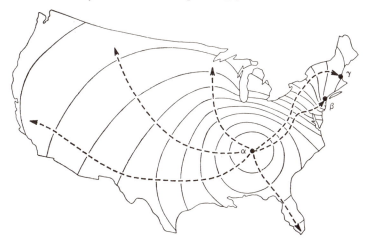

Figure 2.9. Minimum-cost paths and isocost contours for land acquisition from a point (from Haggett and Chorley, 1969; after Warntz, 1965).

on the map from an origin to a destination the normal to the map traces a corresponding path on the surface of the inverse velocity field. The vertical area lying between the path on the map and the path on the surface, bounded by the two vertical lines at the origin and destination of the path, is the total amount of time spent travelling along that path. This area is the integral sum of the elements of time associated with each location on the path.

To obtain the minimum travel time we have to find the minimum vertical area between the map and the inverse velocity field. A path on the plane which corresponds to this minimum area will then be a minimum-time path.

A physical analogue to this problem, which yields the form of the minimum path on the plane, and the time it takes to complete the journey on that path, is the soap bubble model mentioned earlier. The surface representing the inverse velocity field is attached to a flat base (representing the map of the city), and two vertical wires are placed between the surface and the base at the origin and the destination of the trip. This is submerged in a soap solution and withdrawn. A soap film is formed, bounded by the surface, the base, and the two vertical wires. The soap film has the minimum possible surface area. In our case it represents the minimum travel time between the two points. The path traced by the soap film on the base represents the minimum-time path, and the area of the soap film represents the minimum travel time between the origin and the destination.

An isometric projection[3] of a minimum-time path in the inverse velocity field for Greater Manchester is presented in figure 2.10.

In mathematical terms the minimum path from A to B can be found by solving the variational problem of minimising the integral

$$\int_A^B \frac{\mathrm{d}s}{V} \, ,$$

where V is the velocity and $\mathrm{d}s$ the element of distance. The solution to this problem provides us with a differential equation for minimum paths. (see mathematical note 2.1, pp.20–22 for solution of this problem). This equation can be exactly solved for some simple velocity fields. For others, such as the Manchester velocity field, the solution requires numerical approximations.

Let us consider a simple example, the velocity field $V(r) = \omega r$, where ω is some constant. In this field the velocity at the city centre is equal to zero and increases in direct proportion to distance from the centre.

---

[3] The isometric projections were obtained by means of a Calcomp plotter available at the University of London Computer Centre. The subroutine for plotting the surfaces was developed by J. Adams, Royal College of Art, Kensington Gore, London.
The grid in this figure, as in later figures, is drawn at half-mile intervals.

The solution to the variational problem for this field takes the form of a logarithmic spiral. Each spiral cuts, at a constant angle, any straight line passing through the city centre. To find the isochrones to this family we can use the procedure described in mathematical note 2.1. The family of spiral minimum paths and the family of isochrones for a point east of the city centre are illustrated in figure 2.11. The isochrones have been drawn at equal time intervals, except the last one which is the isochrone reached at time $t = \pi/\omega$.

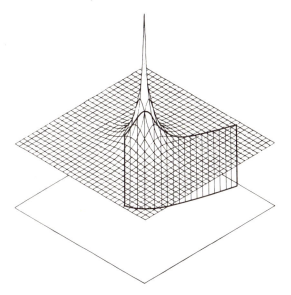

Figure 2.10. A minimum-time path in the inverse velocity field for Greater Manchester, 1965. The vertical area represents total travel time on the path.

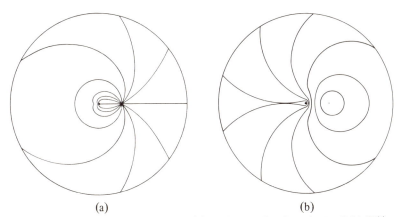

(a)   (b)

Figure 2.11. (a) Minimum paths and (b) isochrones for the velocity field $V(r) = \omega r$.

Let us now consider the problem of constructing the minimum-time paths and isochrones for Greater Manchester. The velocity field for Greater Manchester has been described earlier in this chapter and is illustrated in figure 2.2. When we substitute this velocity field into the differential equation for the minimum path, we obtain an equation with no obvious algebraic solution. We are therefore led to use numerical methods in order to obtain solutions. The numerical solution to the differential equation is incorporated in the computer program described in Angel and Hyman (1971). However, at this stage we are not concerned with detailed numerical results but instead we seek only an overall graphical representation of the family of solutions. We therefore adapt Huygens' construction, described earlier, to obtain the families of minimum-time paths and isochrones from a given origin in the velocity field for Greater Manchester.

To construct isochrones at intervals of Δt we first draw a small circle of radius $V\Delta t$ from the original point on the urban plane. If we regard this as the first isochrone, we then set a compass to have the appropriate radius $V\Delta t$ and from points on the first isochrone we construct a series of arcs. The envelope of these arcs determines the next isochrone. By repeated application of this method the set of isochrones from the original point is constructed. The minimum paths can now be obtained by drawing line segments from a point on one isochrone to the nearest point on the next isochrone. The resulting forms for the minimum-time paths and isochrones are illustrated in figure 2.12 for a point due east of the city centre.

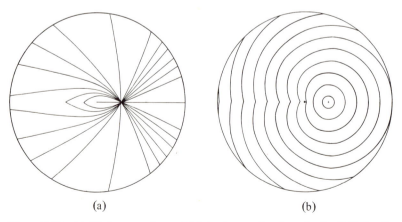

(a)   (b)

Figure 2.12. (a) Minimum paths and (b) isochrones for the velocity field for Greater Manchester, 1965.

2.3 Estimating travel time in radially symmetrical fields

The numerical calculation of travel times is incorporated in the computer program for the spatial interaction model described in the next chapter, but when only a relatively small number of calculations are required, it is possible to obtain travel times by an adaptation of Huygens' construction. The method is used to construct a graph, the *chronograph*, which can be rotated about the city centre to provide estimates for minimum paths, isochrones, and travel times with a fair degree of accuracy. We have used the chronograph for Greater Manchester to test the accuracy of our estimates of travel time, these estimates being compared with estimates obtained by the SELNEC transportation study (1965). The latter were computed from survey data, utilizing a shortest-route algorithm. The velocity field estimates were found to be almost proportional to the network estimates, but in general appeared to be consistently lower (the slope of the regression line is $0 \cdot 74$; the coefficient of linear regression is $0 \cdot 975$). The comparison is illustrated in figure 2.13.

The lower estimates appear to be due to our view of travel as taking place on curved paths, while in reality travellers move through a system of streets and intersections. If this is true then our velocities need to be scaled down by a factor which reflects the resulting increase in journey length. A new velocity field for Greater Manchester may now be obtained by multiplying all velocities by the slope of the line in figure 2.13. This is given by

$$V(r) = 18 \cdot 5 - 12 \cdot 5 \exp(-0 \cdot 56r) \ .$$

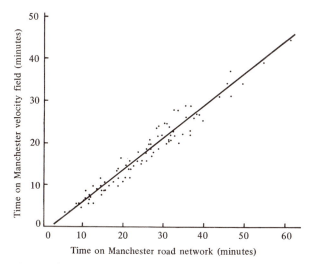

Figure 2.13. A comparison of travel time on the Greater Manchester velocity field with travel time on shortest routes on the road network, 1965.

The chronograph for Greater Manchester was constructed from the set of minimum paths which are orthogonal to a chosen radial route from the city centre. To construct a minimum path from a given point on the radial we draw a circle of radius $V(r)\Delta t$ about this point, where r is the distance of that point from the city centre and Δt is a specified time interval. All points on this circle are Δt from the original point. We draw a set of arcs of radii $V(r)\Delta t$, each centred on a point on the first circle. We then draw the envelope of these arcs. All points on this envelope are at twice the specified time interval from the original point. A further set of arcs is drawn about points of this envelope. By a repeated application of this method we obtain a set of isochrones at intervals Δt from the original point (see figure 2.14a). The minimum path can now be drawn by tracing a curve through the original point, orthogonal to the chosen radial and to each of the isochrones.

A selected set of minimum paths, orthogonal to this radial, is drawn by repeating this procedure for various points on the radial (2.14b). The time contours can now be drawn by connecting the points of intersection of each minimum path with the appropriate isochrone (2.14c). The time contour will then describe the set of points which are a fixed travel time from the original radial. The chronograph for the Greater Manchester velocity field is presented in figure 2.15.

This particular chronograph has the property that the minimum paths orthogonal to the original radial reach a limiting angle as measured by their deviation from a straight line. This angle is reached by the path whose minimum radius is $0 \cdot 56$ miles. The paths then begin to straighten out again. The minimum path into the city centre is thus a straight line orthogonal to the original radial. Travel time for trips on a path which

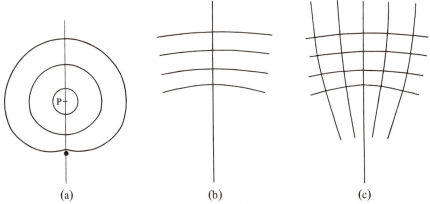

(a)   (b)   (c)

Figure 2.14. Construction of the chronograph. (a) A set of isochrones. Points on the isochrones are $n\Delta t$ (where n is an integer) from the point P marked on the radial. (b) Set of minimum paths, drawn through, and orthogonal to, the radial. (c) Minimum paths intersected by time contours.

comes within 0·56 miles from the centre must therefore be measured using the dotted lines in figure 2.15. To obtain readings of travel times between two points, we rotate the chronograph about the centre of the city until the two points lie on one minimum path in the top half of the chronograph, or in a band between two minimum paths. We then find the sum or the difference of the times from the original radial to each of the points to obtain the travel time (depending on whether the radius lies between the two points or to one side of them).

A chronograph, drawn by the method described above, was used in estimating the travel times on the Manchester velocity field—these are presented in figure 2.13. If one uses the procedure described in mathematical note 2.1 it is possible to construct a more accurate chronograph with the aid of a computer. The chronograph drawn in figure 2.15 was constructed in such a manner.

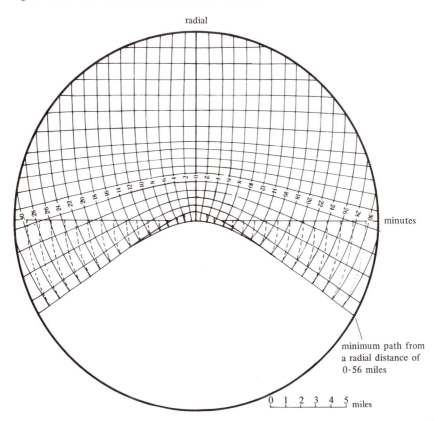

Figure 2.15. The chronograph for Greater Manchester, 1965. Time contours are drawn at two minute intervals from the original radial.

We now have a framework for describing and analysing movement in urban areas without recourse to network procedures. The mathematical tools required for using this framework are presented in the note which follows, and in later chapters we shall employ the framework to discuss a series of geographical problems where movement plays a key role.

Mathematical note 2.1. *A procedure for calculating minimum paths, isochrones, and travel time in a velocity field*
Given a velocity field, we wish to obtain mathematical expressions for the form of minimum-time paths between any two points, the value of travel time between these points, and the set of isochrones associated with each point. The analysis also applies to cost fields and travel expenditures, as all the expressions can be converted to cost instead of time. For convenience the discussion will be structured in the form of definitions, theorems, and proofs; worked examples will accompany the analysis to illustrate the various concepts.

Consider a polar coordinate system (r, θ), see figure 3.10, p.45, and a velocity field, $V = V(r)$. An element of distance Δs on the plane is, in polar coordinates (r, θ),

$$\Delta s = (\Delta r^2 + r^2 \Delta \theta^2)^{1/2} . \tag{2.1}$$

An element of time Δt can be expressed in the form

$$\frac{(\Delta r^2 + r^2 \Delta \theta^2)^{1/2}}{V} . \tag{2.2}$$

The travel time on a path P from (r_1, θ_1) to (r_2, θ_2) is thus given by:

$$t(r_1, \theta_1, r_2, \theta_2) = \int_P \frac{1}{V} \left[\left(\frac{dr}{d\theta} \right)^2 + r^2 \right]^{1/2} d\theta . \tag{2.3}$$

Definition: Between two points on the plane the path which can be traversed in the smallest time is said to be the *minimum-time path*.

Lemma 2.1. Minimum-time paths from the city centre are radii whenever the velocity field is radially symmetrical.

Proof. In this case we rewrite the integral of equation (2.3) in the form

$$\int_0^{r_1} \frac{1}{V} \left[1 + r^2 \left(\frac{d\theta}{dr} \right)^2 \right]^{1/2} dr \tag{2.4}$$

for a path $\theta = \theta(r)$ between the centre and a point (r_1, θ_1). For a radial path from the centre to (r_1, θ_1) we have

$$\frac{d\theta}{dr} = 0 . \tag{2.5}$$

The time along that path is

$$\int_0^{r_1} \frac{\mathrm{d}r}{V} \, , \tag{2.6}$$

and since

$$\left[1 + r^2 \left(\frac{\mathrm{d}\theta}{\mathrm{d}r} \right)^2 \right]^{1/2} \geq 1 \, , \tag{2.7}$$

and the velocity of travel is a function of the radius only, the value of the integral in equation (2.4) must always exceed the value of the integral in equation (2.6). Hence the minimum paths from the centre are radii.

Theorem 2.1. A minimum-time path satisfies the differential equation

$$\frac{\mathrm{d}r}{\mathrm{d}\theta} = \frac{r}{KV} (r^2 - K^2 V^2)^{1/2} \, , \tag{2.8}$$

where K is a constant associated with the path, whenever the velocity field is radially symmetrical.

Proof. Travel time on any path P is given by equation (2.3). To obtain a minimum-time path, the integral of equation (2.3) must be minimized. To simplify the notation we write this problem in the form

$$\text{minimize} \int F\left(r, \frac{\mathrm{d}r}{\mathrm{d}\theta} \right) \mathrm{d}\theta \tag{2.9}$$

where

$$F\left(r, \frac{\mathrm{d}r}{\mathrm{d}\theta} \right) = \frac{1}{V} \left[r^2 + \left(\frac{\mathrm{d}r}{\mathrm{d}\theta} \right)^2 \right]^{1/2} . \tag{2.10}$$

The solution to this variational problem must satisfy Euler's differential equation. [The conditions for the existence of the minimum are discussed and presented in Courant and Hilbert (1953, p.184).]

$$\frac{\mathrm{d}}{\mathrm{d}\theta} \left(\frac{\partial F}{\partial r'} \right) = \frac{\partial F}{\partial r} \, , \tag{2.11}$$

where $r' = \mathrm{d}r/\mathrm{d}\theta$. Since F does not involve θ, we observe that

$$\frac{\mathrm{d}}{\mathrm{d}\theta} \left(F - r' \frac{\partial F}{\partial r'} \right) = r' \frac{\partial F}{\partial r} + r'' \frac{\partial F}{\partial r'} - r'' \frac{\partial F}{\partial r'} - r' \left[\frac{\mathrm{d}}{\mathrm{d}\theta} \left(\frac{\partial F}{\partial r'} \right) \right] = r' \left[\frac{\partial F}{\partial r} - \frac{\mathrm{d}}{\mathrm{d}\theta} \left(\frac{\partial F}{\partial r'} \right) \right] . \tag{2.12}$$

The above expression vanishes by Euler's equation and so we can write (Courant and Hilbert, 1953, p.206)

$$F - r' \frac{\partial F}{\partial r'} = K \, , \tag{2.13}$$

where K is a constant of integration.

If we substitute for F from equation (2.10), and rewrite r' as $dr/d\theta$, we obtain the differential equation of the minimum-time paths

$$\frac{dr}{d\theta} = \frac{r}{KV}(r^2 - K^2V^2)^{\frac{1}{2}} \,. \tag{2.14}$$

The reader should note that in the general radially nonsymmetrical case the simplification introduced here does not apply. To obtain the minimum paths we must solve equation (2.11) directly, where F depends on both r and θ. This is more difficult than in the radially symmetrical case since equation (2.11) is a second-order differential equation. It is possible to obtain a first-order equation when the velocity field is a function of one cartesian coordinate only, say $V = V(x)$. In this case it can be verified that the differential equation of minimum paths is given by

$$\frac{dy}{dx} = \frac{KV}{(1 - K^2V^2)^{\frac{1}{2}}} \,.$$

This particular form of the velocity field, $V = V(x)$, can be applied to situations where there is one linear corridor of fast movement, and speeds fall off gradually as one moves away from this corridor—the type of field that could be associated with a motorway. This particular type of case requires further study. In the remainder of this note, however, it will be assumed that the velocity field is radially symmetrical.

Definitions. A *route* is a minimum path which is traversed in a specified direction. The *extension* of a route between two points is the union of all routes passing through the two points. The constant of integration K of equation (2.13) is said to be the *characteristic* of a route. K is also the characteristic of the extension of a route.

Theorem 2.2. Two routes have the same characteristic if and only if their extensions may be transformed into each other by a rotation about the centre.

Proof. If two routes have the same characteristic K then, for each value of the radius r, their extensions must have the same value for $d\theta/dr$ [equation (2.8)]. So for each radius the values of θ for the two extensions must differ by a constant. Thus one extension may be transformed into the other by a rotation about the centre. Conversely if such a transformation exists, then for each radius r the two extensions must have the same value for $r(r^2 - K^2V^2)^{\frac{1}{2}}/KV$. Hence they must have the same characteristic K, and so the original routes must have the same characteristic.

Definition. The *minimum radius*, $r_{\min}(K)$, of a route with characteristic K is the shortest distance from the centre of the city to the extension of the route.

This is illustrated in the figure 2.16.

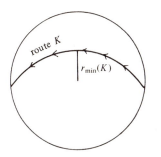

Figure 2.16. The minimum radius $r_{\min}(K)$ of a route with characteristic K.

Theorem 2.3. For velocity fields $V(r)$, with the properties that $r/V(r)$ is a monotonically increasing function of r and

$$\lim_{r \to 0} \frac{r}{V(r)} = 0 \,, \tag{2.14}$$

the equation

$$r = |K|V(r) \tag{2.15}$$

has the unique solution $r_{\min}(K)$ for each characteristic K.

Proof. Suppose K is the characteristic of a route whose extension does not pass through the centre. This extension must have a point of closest approach to the centre. At this point $(dr/d\theta) = 0$. Solving equation (2.8) for K, when $(dr/d\theta) = 0$, yields

$$K = \pm\frac{r}{V} \,, \tag{2.16}$$

and hence equation (2.15) is satisfied. Thus $r_{\min}(K)$ is a solution to equation (2.15). Since $r/V(r)$ increases monotonically with r this solution must be unique. If the extension passes through the centre, it must have zero characteristic by virtue of equation (2.14). So equation (2.15) will have $r = 0$ as a unique solution[4].

---

[4] The analytical procedure is restricted by the conditions of theorem 2.3 to those velocity fields which satisfy the requirement that $r/V(r)$ be a monotonically increasing function of r. This condition can be expressed in the form

$$\frac{V}{r} > \frac{dV}{dr} \qquad \text{for} \quad 0 < r < \infty \,.$$

This requirement is satisfied for the Greater Manchester velocity field and will probably be met by most forms of velocity fields computed in existing urban areas.

Theorem 2.4. A minimum-time path with characteristic K satisfies the equation

$$\frac{r}{V}\sin\psi = K,$$ (2.17)

where ψ is the angle between the path and the radius.

From figure 2.17 we have

$$\sin\psi = r\frac{d\theta}{ds} = r\Big/\left[r^2+\left(\frac{dr}{d\theta}\right)^2\right]^{1/2} = 1\Big/\left[1+\frac{1}{r^2}\left(\frac{dr}{d\theta}\right)^2\right]^{1/2}.$$ (2.18)

From theorem 2.1 we obtain

$$\frac{1}{r^2}\left(\frac{dr}{d\theta}\right)^2 = \frac{r^2}{K^2V^2}-1.$$ (2.19)

By combining equations (2.18) and (2.19) we obtain

$$\sin\psi = \frac{1}{(r^2/K^2V^2)^{1/2}} = \frac{KV}{r},$$ (2.20)

and hence derive the required result.

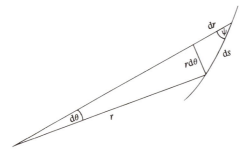

Figure 2.17.

Definition. The *maximal characteristic*, $\overline{K}(r_1,r_2)$, is the positive characteristic of a route whose minimum radius is the smaller of r_1 and r_2.

By theorem 2.3, we have

$$\overline{K}(r_1,r_2) = \frac{\min(r_1,r_2)}{V[\min(r_1,r_2)]}.$$ (2.21)

Definition. The *critical angle*, $\overline{\theta}(r_1,r_2)$, is the positive angular difference of a route between two radii r_1 and r_2 with a maximal characteristic $\overline{K}(r_1,r_2)$. The maximal characteristic and the critical angle are illustrated in the figure 2.18.

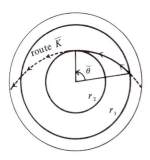

Figure 2.18. The maximal characteristic $\overline{K}(r_1,r_2)$ and the critical angle $\overline{\theta}(r_1,r_2)$.

Lemma 2.2. The relationship between the critical angle and the maximal characteristic of a given pair of radii, r_1 and r_2, is given by

$$\overline{\theta}(r_1,r_2) = \overline{K}(r_1,r_2) \left| \int_{r_1}^{r_2} \frac{V\,dr}{r[r^2 - \overline{K}^2(r_1,r_2)\,V^2]^{\frac{1}{2}}} \right| . \tag{2.22}$$

Proof. Equation (2.8) gives us

$$\frac{d\theta}{dr} = \frac{KV}{r(r^2 - K^2V^2)^{\frac{1}{2}}} . \tag{2.23}$$

We substitute $\overline{\theta}(r_1,r_2)$ for θ and $\overline{K}(r_1,r_2)$ for K, and integrate this expression between the two radii to obtain the desired result.

Definition. The *angular difference*, θ_{12}, is the angle traversed by the route from an origin (r_1,θ_1) to a destination (r_2,θ_2).

θ_{12} may be calculated from the following expressions:

$$\theta_{12} = \theta_2 - \theta_1 \qquad \text{for} \quad -\pi \leqslant \theta_2 - \theta_1 \leqslant \pi , \tag{2.24}$$

$$\theta_{12} = \theta_2 - \theta_1 - 2\pi \qquad \text{for} \quad \pi < \theta_2 - \theta_1 \leqslant 2\pi ,$$

$$\theta_{12} = \theta_2 - \theta_1 + 2\pi \qquad \text{for} \quad -2\pi \leqslant \theta_2 - \theta_1 < -\pi .$$

θ_{12} is restricted to the range $-\pi \leqslant \theta_{12} \leqslant \pi$.

Definition. A *direct route* is a route which does not pass through its minimum radius. A *through route* is a route which passes through its minimum radius.

Note that the route between two points is a direct route if and only if the absolute value of their angular difference is smaller than the critical angle corresponding to their radii.

Theorem 2.5. The relationship between the angular difference θ_{12} and the characteristic K of a direct route between two points is given by

$$\theta_{12} = \int_{r_1}^{r_2} \frac{KV\,dr}{r(r^2 - K^2V^2)^{\frac{1}{2}}} \qquad \text{when} \quad r_1 < r_2 , \tag{2.25}$$

and

$$\theta_{12} = \int_{r_2}^{r_1} \frac{KV\mathrm{d}r}{r(r^2 - K^2V^2)^{\frac{1}{2}}} \qquad \text{when} \quad r_1 > r_2 \,. \tag{2.26}$$

For a through route the relationship is given by

$$\theta_{12} = \int_{r_{\min}(K)}^{r_1} \frac{KV\mathrm{d}r}{r(r^2 - K^2V^2)^{\frac{1}{2}}} + \int_{r_{\min}(K)}^{r_2} \frac{KV\mathrm{d}r}{r(r^2 - K^2V^2)^{\frac{1}{2}}} \,. \tag{2.27}$$

Proof. If we substitute θ_{12} for θ in equation (2.23) and integrate this expression between the two radii as limits we obtain the desired result. Since the integration is in the positive direction, from the smaller to the larger radius, the sign of θ_{12} will be the same as that of K. It can be verified that K is positive for anticlockwise routes and negative for clockwise routes. θ_{12} is positive for anticlockwise routes and negative for clockwise routes by definition.

Corollary 2.1. The equation of a minimum-time path, $\theta = \theta(r)$, for a direct route between two points (r_1, θ_1) and (r_2, θ_2) is given by

$$\theta(r) = \theta_1 + 2k\pi + \int_{r_1}^{r} \frac{KV\mathrm{d}u}{u(u^2 - K^2V^2)^{\frac{1}{2}}} \qquad \text{when} \quad r_1 < r_2 \,, \tag{2.28}$$

and

$$\theta(r) = \theta_1 + 2k\pi + \int_{r_2}^{r} \frac{KV\mathrm{d}u}{u(u^2 - K^2V^2)^{\frac{1}{2}}} \qquad \text{when} \quad r_2 < r_1 \,. \tag{2.29}$$

For a through route, the equation of a minimum-time path is given by

$$\theta(r) = \overline{\theta}\,[r_1, r_{\min}(K)] + 2k\pi \int_{r_{\min}(K)}^{r} \frac{KV\mathrm{d}u}{u(r^2 - K^2V^2)^{\frac{1}{2}}} \,, \tag{2.30}$$

where $\overline{\theta}\,[r_1, r_{\min}(K)]$ is the critical angle between r_1 and $r_{\min}(K)$, and K is evaluated from the equations of theorem 2.5; k takes on the values -1, 0, or $+1$ to ensure that $\theta(r)$ is within the range 0 to 2π.

Proof. By replacement of θ_2 by θ in equations (2.24), and of r_2 by r in equations (2.25), (2.26), and (2.27), and substitution of the appropriate equation for the angular difference, we obtain the desired results. In the case of a through route, the equation of the path is obtained by integrating from the minimum radius of the path. The initial angle at that minimum radius is therefore the critical angle between that radius and r_1.

The above corollary makes it possible to compute the equation of minimum-time paths between any pair of points in a radially symmetrical velocity field. Given the coordinates of the two points we can compute their angular difference from the definition. We can then compute their maximal characteristic by means of equation (2.21), and the critical angle of the route joining them by means of lemma 2.2. Comparison of the angular difference and the critical angle enables us to decide whether the route is a direct route or a through route. We can then find the characteristic of the route by using the equations of theorem 2.5, and obtain the equation of the path by means of corollary 2.1.

Theorem 2.6. Travel time on a route with characteristic K satisfies the differential equation

$$\frac{dt}{dr} = \frac{r}{V(r^2 - K^2 V^2)^{\frac{1}{2}}} \cdot \tag{2.31}$$

Proof. From equation (2.2) we obtain the following expression by dividing by Δr and taking the limit as $\Delta r \to 0$:

$$\frac{dt}{dr} = \left[1 + r^2 \left(\frac{d\theta}{dr}\right)^2\right]^{\frac{1}{2}} V . \tag{2.32}$$

On a route with characteristic K we have

$$\frac{d\theta}{dr} = \frac{KV}{r(r^2 - K^2 V^2)^{\frac{1}{2}}} \cdot \tag{2.23}$$

On substituting this expression for $d\theta/dr$ in equation (2.28) we obtain the following expression for travel time on a route with characteristic K:

$$\frac{dt}{dr} = \frac{r}{V(r^2 - K^2 V^2)^{\frac{1}{2}}} \cdot$$

Theorem 2.7. Minimum travel time between two points with radial coordinates r_1 and r_2 on a direct route with characteristic K is given by

$$t_D(r_1, r_2, K) = \int_{r_1}^{r_2} \frac{r dr}{V(r^2 - K^2 V^2)^{\frac{1}{2}}} \qquad \text{for} \quad r_1 < r_2 , \tag{2.33}$$

$$t_D(r_1, r_2, K) = \int_{r_2}^{r_1} \frac{r dr}{V(r^2 - K^2 V^2)^{\frac{1}{2}}} \qquad \text{for} \quad r_1 > r_2 , \tag{2.34}$$

and on a through route by

$$t_T(t_1, r_2, K) = \int_{r_{min}(K)}^{r_1} \frac{r dr}{V(r^2 - K^2 V^2)^{\frac{1}{2}}} + \int_{r_{min}(K)}^{r_2} \frac{r dr}{V(r^2 - K^2 V^2)^{\frac{1}{2}}} \cdot \tag{2.35}$$

Proof. Integration of equation (2.31) within the appropriate limits yields the desired results. Since the expression in equation (2.31) is always positive, the integral of this expression between the smaller and the larger radii will also always be positive. Hence travel time will always be positive, as required.

Theorem 2.7 allows us to calculate the value of the minimum travel time between any pair of points in a velocity field. Given such a pair of points, we first find the characteristic of the minimum-time path between them, as described earlier, and this characteristic is then used to obtain its travel times from the appropriate equation in theorem 2.7. Let us use a worked example to illustrate better to the reader the whole procedure.

Consider the velocity field

$$V(r) = \omega r ,$$

where ω is some constant. By substituting for V in equation (2.8) we obtain

$$\frac{dr}{d\theta} = \frac{r}{K\omega}(1 - K^2\omega^2)^{\frac{1}{2}} = mr , \tag{2.36}$$

where

$$m = \frac{(1 - K^2\omega^2)^{\frac{1}{2}}}{K\omega} . \tag{2.37}$$

Equation (2.36) is a first-order differential equation and its solution yields

$$\ln r = m\theta + C , \tag{2.38}$$

where both constants, m and C, can be evaluated for a given minimum-time path between any pair of points (r_1, θ_1) and (r_2, θ_2). The reader may recognize that equation (2.38) is an equation for a logarithmic spiral which cuts each radius at a constant angle. By setting $m = 0$ we obtain the minimum path, which is a circle $r = e^C$. The family of minimum paths for a given point is demonstrated in figure 2.11 (a). Since each spiral minimum route reaches the city centre, it does not possess a minimum radius. This means the conditions of theorem 2.3 are not satisfied in this case since

$$\frac{r}{V(r)} = \frac{1}{\omega} = \text{constant} , \tag{2.39}$$

and therefore we need not calculate a maximal characteristic or a critical angle, since there are no through routes.

To obtain the travel time between the two points we first compute their angular difference θ_{12} from equation (2.24)—we assume $r_1 < r_2$. Equation (2.25) takes the form

$$\theta_{12} = \frac{K\omega}{(1 - K^2\omega^2)^{\frac{1}{2}}} \int_{r_1}^{r_2} \frac{dr}{r} = \frac{1}{(1 - K^2\omega^2)^{\frac{1}{2}}} \ln\left(\frac{r_2}{r_1}\right) . \tag{2.40}$$

We can now use this equation to find an expression for the travel time between the two points from equation (2.33),

$$t_D(r_1, r_2, K) = \frac{1}{\omega(1 - K^2\omega^2)^{\frac{1}{2}}} \int_{r_1}^{r_2} \frac{dr}{r} = \frac{1}{\omega(1 - K^2\omega^2)^{\frac{1}{2}}} \ln\left(\frac{r_2}{r_1}\right) . \tag{2.41}$$

By substituting for K from equation (2.40) we get

$$t_D(r_1, r_2, \theta_{12}) = \frac{1}{\omega}\left[\ln\left(\frac{r_2}{r_1}\right)^2 + \theta_{12}^2\right]^{\frac{1}{2}} . \tag{2.42}$$

The reader should notice that this equation holds also for $r_1 \geqslant r_2$, and thus it provides a formula for computing the travel time between any pair of points in this field.

To find the equation of an isochrone, at time t_0, from a given point (r_1, θ_1) we must solve for θ_{12} in equation (2.42), given a value for t_0. This solution yields

$$\theta_{1r} = \left[\omega^2 t_0^2 + \ln \left(\frac{r}{r_1} \right)^2 \right]^{1/2}, \tag{2.43}$$

where θ_{1r} is the angular difference between the two points (r_1, θ_1) and (r, θ). The latter equation gives two values for the angular difference for any given value of r. These values are equal in magnitude but of opposite sign. Thus the isochrones are found to be symmetrical about the radial connecting the given point (r_1, θ_1) with the city centre.

The isochrone at $t_0 = \pi/\omega$ is tangential to itself at $(r_1, -\theta_1)$, as can be seen by examination of equation (2.43). The family of isochrones for a given point is illustrated finally in figure 2.11 (b).

Transformations and geographical theory

3.1 Euclidean geography

In the previous chapter we characterized travel as taking place along curved paths on a flat surface. We refer to this surface as the *Euclidean plane*. The Euclidean plane forms the basic geometrical framework for the construction of maps by geographers and cartographers. The application of this framework in geography is typified by the early attempts to represent the world as a flat surface, as in the example depicted in figure 3.1—the Ebsdorfer world map, assembled by a German monk in the eleventh century.

Figure 3.1. The Ebsdorfer map of the world (*circa* 11th century). Jerusalem is shown at the centre with the continents of Asia, Europe, and Africa extending from it toward the external sea.

Later map making has been concentrated on producing various projections of the spherical surface of the earth into the Euclidean plane. The projections implicit in the early medieval maps have been analyzed by Tobler (1966).

The shortest paths on any map which has been produced by a projection of a spherical surface into the plane cannot be straight lines. In fact, since the earth is spherical, it is impossible to construct a flat map of the world which correctly represents distances. For a similar reason it is not possible to construct a flat map of the city which correctly represents travel time, as we shall see later.

The shortest paths on a curved surface, such as the earth's spherical surface, are referred to as *geodesics*.

"Time geodesics or cost geodesics, although generally not straight paths over the earth's physical surface, are nevertheless straight with respect to the surfaces on which they lie. However much they may bend and turn with respect to the earth's surface as the datum, they do not turn at all but rather move only straight forward with respect to the surface in whose terms they are defined" (Warntz, 1967, p.8—page 108 of this book).

A familiar example of geodesic paths is provided by the great circles, such as the circles of longitude, on the spherical surface of the earth. A polar projection to the Euclidean plane can be constructed which maps the circles of longitude into straight lines, emanating from the image of the pole. However the great circle between two points, having different longitudes will not be represented by a straight line on the transformed plane.

Instead of representing travel as taking place on curved paths in the urban plane, we shall employ transformations from the plane to a curved surface. This transformation has the property that the time taken to traverse a path in the urban plane is equal to the length of the image of this path on the curved surface. So it has the particular property that minimum time paths in the urban plane are represented by geodesics on the curved surface. This surface is referred to as the *time surface*.

This surface is not to be confused with the inverse velocity field, defined in the previous chapter, where travel time was represented by an area, rather than by the length of a path.

In this chapter we shall employ transformations to the time surface in order to critically examine the spatial frameworks postulated by theories in regional science and geography. These theories frequently employ simplifying assumptions in order to use Euclidean geometry as a spatial framework. We consider in particular the works of von Thünen (1966, first published 1875), Christaller (1966, first published 1933), and Lösch (1967, first published 1940), which have laid the foundations for theoretical studies of regional geographical space.

The purpose of this chapter is to look closely at the idea, which has often been expressed in the literature, that geographical space, with its nonuniform distribution of transport costs, can be transformed into some abstract space where some or all of the simplifying assumptions of these theories hold.

The simplifying assumptions are quite explicit. In von Thünen, for example, we find the following statement:

"Imagine a very large town, at the centre of a fertile plain, which is crossed by no navigable river or canal. Throughout the plain the soil is capable of cultivation and of the same fertility. Far from the town, The plain turns into an uncultivated wilderness which cuts off all communication between this State and the outside world" (von Thünen, 1966, p.7).

Given that the market for all agricultural produce is in the town, and that transport cost is proportional to distance, von Thünen derives a pattern of agricultural production. The crops yielding the highest returns per unit of land are in a ring closest to the market centre. Crops yielding lower returns are produced in rings further from the market. These rings are illustrated in figure 3.2.

The bottom half of the figure is an attempt to illustrate a more realistic distribution of agricultural production, but the author makes little attempt to explain how his results could be applied to such a distribution. Christaller and Lösch make similar assumptions to those of von Thünen in their representation of geographical space, without assuming a single centre. By means of further economic assumptions Christaller and Lösch derive a pattern for the location of production centres. Towns of a given size are uniformly distributed in space, each having a hexagonal market area.

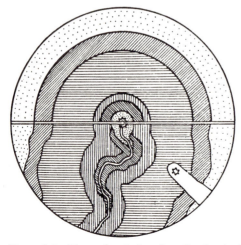

Figure 3.2. Rings of agricultural production in the Isolated State (from von Thünen, 1966).

These sets of hexagonal market areas, for towns of different sizes, are superimposed on each other. This is illustrated in figure 3.3.

We can characterize the spatial framework of these authors by means of three basic assumptions. The first is that the terrain under study is flat and uninterrupted. We shall refer to this as *the assumption of the Euclidean plane*. The second is that important quantities on the terrain are uniformly distributed throughout. These quantities might be resource endowments, agricultural fertility, purchasing power, population density, and the like. We shall refer to this as *the assumption of uniform densities*. The third is the assumption that movement can proceed with equal ease everywhere, regardless of location or direction of travel. We shall refer to this as *the assumption of uniform transport facility*.

These assumptions have come under severe criticisms. To early geometers the only available measure of distance was that defined by Euclidean geometry. The Euclidean metric was therefore regarded as the absolute measure of distance. Recently the applicability of Euclidean geometry to geographical space has been challenged.

"Distance, it seems, can be measured only in terms of process and activity. There is no independent metric to which all activity can be referred. In the discussion of the diffusion of information, distance is measured in terms of social interaction; in the study of migration, distance may be measured in terms of intervening opportunity, and so on" (Harvey, 1969, p.210).

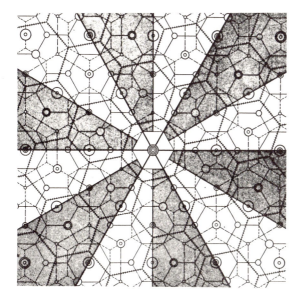

Figure 3.3. Market areas in the Löschian landscape (from Isard, 1956; after Lösch).

Harvey thus makes two important observations. First, that the intrinsic metric of the Euclidean plane—the straight line distance—is not an absolute measure of distance. Second, that different phenomena need to be studied by means of different metrics. Harvey's criticism seems all the more appropriate in view of the lack of correspondence between the predictions of these theories and the observed geographical patterns.

The continued application of Euclidean geometry, however, is seen to be of vital importance in regional geographical theory. It has been postulated, therefore, that there are transformations which make it possible to represent the metrics, or densities, corresponding to different activities by distances or areas on the Euclidean plane. Unless such transformations exist, the regional theories requiring Euclidean geometry cannot be supported by empirical evidence. The belief that the spatial assumptions mentioned earlier can be retained for purposes of analysis by means of transformations is shared by many writers. For example,

"The familiar assumptions of location theory (e.g. flat plane surfaces, equal transport facility in all directions, and uniform resource endowment) are assumptions which are specifically designed to allow a Euclidean treatment of the problem. The solution of the problem may later be disturbed to take account of non-Euclidean properties of actual spaces" (Harvey, 1969, p.225).

"... we should not expect regular hexagonal territories to be generally visible on the earth's surface, because they are related not to geographical space but to population of income space. Hexagons may therefore be thought to be *latent* in most human organization but only through appropriate transformations of geographical space is their form likely to be made visible" (Haggett, 1965, p.55).

This approach has intrigued geographers for many years. In this chapter we hope to shed light on its possibilities and limitations.

3.2 Geometrical transformations

By analogy to the established work on geometrical transformations in cartography, several attempts have been made to find transformations which will bring into life the patterns postulated by von Thünen, Christaller, Lösch, and other regional theorists. Bunge (1964, pp.20–21) poses the problem as follows:

"The difficulty in testing the worth of Christaller and other theoretical geographers has always been that we cannot find an area of the earth's surface which is not distorted by transportation routes and/or disuniform rural population. Some geographers have incorrectly claimed this initial difficulty is grounds for abandoning Christaller altogether while others have correctly seen it as a challenge to find ways to 'straighten out' the map so that the transportation (morphological) and density effects were removed".

A solution to this problem would be to find a transformation of the Euclidean plane, with a realistic pattern of densities and a realistic distribution of costs of travel, into a Euclidean plane with uniform densities and uniform transport facility. Various authors have made limited attempts at finding such a transformation. In each case an attempt has been made to transform the existing geography in such a way that one or two of the assumptions mentioned earlier will be satisfied. These attempts are presented here as four conjectures, each named after the author associated with it.

The first conjecture is *Tobler's conjecture*: A transformation exists which will map a realistic density on the Euclidean plane into a uniform density on the Euclidean plane. Tobler thus attempts to maintain the assumptions of uniform densities and the Euclidean plane. He indeed provides the required construction. Consequently it might be more appropriate to refer to the conjecture as a theorem. A more detailed discussion of his work is presented in section 3.3 below.

The second conjecture is *Warntz's conjecture*: A transformation exists which will map a realistic distribution of transport costs on the Euclidean plane into a curved surface with uniform transport costs. Warntz thus attempts to maintain the assumption of uniform transport facility. In section 3.4 we show that this conjecture is true, and present such a transformation for various velocity fields. The properties of this transformation allow us to shed light on the remaining conjectures.

The third conjecture is *Wardrop's conjecture*: A transformation exists which will map a realistic distribution of transport costs on the Euclidean plane into a Euclidean plane with a uniform distribution of transport costs. Wardrop thus attempts to maintain both the assumption of uniform transport facility *and* the Euclidean plane. In section 3.5 we show that this conjecture is false. A surface obtained by smoothing out transport cost variations is, in general, curved.

The fourth conjecture is *Bunge's conjecture*: A transformation exists which will map a realistic density and a realistic distribution of travel costs into a curved surface with uniform density and a uniform distribution of travel costs. Bunge thus attempts to maintain the assumptions of uniform density *and* equal transport facility. In section 3.6 we show that this conjecture is also false. In the light of these results, we examine the implications for regional and geographical theories.

3.3 Tobler's conjecture

A transformation exists which will map a realistic density on the Euclidean plane into a uniform density on the Euclidean plane.

An early attempt to modify the Löschian analysis by taking into account variations in the density of the consuming population was made by Isard (1956, pp.271–274). Isard points out an inconsistency in the Löschian construction. This construction yields different sizes of concentrations of industrial activity in different locations, and consequently implies nonuniform distributions of the labour force and of the consuming population. The population density is higher in the larger urban concentrations. In these locations, the size of market areas needs to be smaller if each market area is to contain an equal number of consumers. Isard attempts to illustrate the modification of the hexagonal pattern due to these density variations, and presents a pattern of distorted hexagons which increase in size with distance from a major centre. Isard's illustration is presented in figure 3.4.

Tobler's conjecture is a more explicit statement of the problem (Tobler, 1963—this book page 135). He postulates that the distribution of some geographical phenomena, say the population in an urban region, is described by a density function over the Euclidean plane. The total population in the region under consideration is given as an integral expression. In particular the population of any subregion is given by the integral of the density over that subregion. This integral is equated to the area of a transformed subregion in another Euclidean plane. When this condition is satisfied for all subregions, Tobler obtains a set of differential equations. These equations provide sufficient conditions for the transformation to correctly represent the population density as a uniform density on the transformed plane. Tobler points out that the transformations thus obtained are generalizations of equal-area projections.

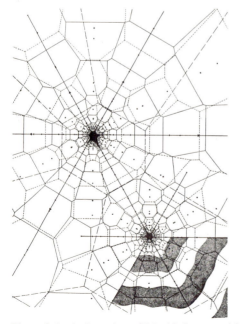

Figure 3.4. A distortion of Lösch's hexagonal lattices to account for density variations (from Isard, 1956).

He further remarks that there is an infinite number of solutions to this problem for any given density, so that additional conditions on the transformations may be imposed. In particular the transformation can be constructed so that travel cost from the map centre is represented by distance to the centre of the transformed map. This form of transformation is applied by Tobler to generalize von Thünen's theory in order to deal with a variable pattern of agricultural fertility and any distribution of transport costs which depends only on distance from the centre. Tobler raises the problem of correctly representing a general, radially non-symmetric distribution of travel costs to the centre. Such a transformation has still to be constructed.

A transformation which illustrates how a nonuniform density is mapped into a uniform density is illustrated in figure 3.5. The top diagram describes an area of nonuniform rural population density on a Euclidean plane with a regular rectangular grid. The bottom diagram represents a

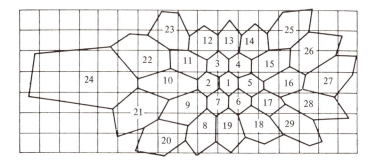

(a)

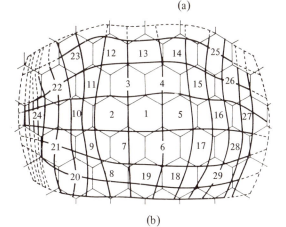

(b)

Figure 3.5. (a) A nonuniform rural population density with a superimposed approximation of a distorted hexagonal lattice; (b) a transformed map with even population densities with a superimposed regular hexagonal lattice (from Harvey, 1969; after Bunge, 1966).

transformed plane on which the density is uniform, and the rectangular grid is distorted. Tobler's theorem makes it possible to employ the Euclidean plane in theories which postulate the assumption of uniform density. The following section will discuss the theories which postulate the assumption of uniform transport facility.

3.4 Warntz's conjecture

A transformation exists which will map a realistic distribution of transport costs on the Euclidean plane into a curved surface with uniform transport facility.

The assumption of uniform transport facility is postulated in many geographical theories of location, where the distance between points is assumed to be the length of the shortest path connecting the points. When travel costs vary over a geographical space, minimum-cost paths deviate from shortest paths, as shown earlier (chapter 2). Warntz seeks a representation of space in which these minimum-cost paths correspond to shortest paths on a curved surface. He postulates the existence of such a representation (Warntz, 1967, pp.7–8):

> "For the sake of analysis, we can imagine a portion of the physical surface of the earth as transformed into a time surface or a cost surface, for example, with optimum paths regarded as geodesics on their appropriate surfaces".

Given a velocity field on the Euclidean plane, we define a transformation of the plane into a two-dimensional curved surface lying in three-dimensional Euclidean space. The surface characterized by the transformation has the property that travel time on any path in the original Euclidean plane is equal to the length of the image of that path on the transformed surface. In particular, the image of the minimum-time path between two points on the plane is the geodesic curve joining their image points on the surface. This surface has therefore been referred to as the time surface.

The transformation to the time surface thus has the required property that distance on this surface represents travel time. The time spent travelling a unit distance is clearly constant on this surface. In mathematical note 3.1 we obtain a transformation with the above property for a large class of radially symmetric fields. This transformation is used in the derivation of several examples of time surfaces, which are presented below. The transformation is smooth and conformal, that is all its derivatives exist, and it preserves angles between curves.

The transformation of the field into a time surface provides us with another method for obtaining the forms of minimum-time paths, and isochrones, for any given point on the original Euclidean plane. This plane is first transformed into a time surface associated with its given velocity field. The set of geodesics through the image of the original point is constructed. These geodesics are then transformed back into the plane, and the required family of minimum paths is obtained. The family of isochrones of the original point can be obtained by transforming the family of contours on the time surface, which are at a given distance from the image of the original point, to the original plane. Travel time between points on the plane can be obtained by measuring the distance between the images of these points on the time surface.

In the following paragraphs we illustrate the transformation of the field into a time surface with several examples. Consider again the velocity field $V(r) = \omega r$, where ω is some constant. The time surface for this velocity field is shown to be an infinitely long cylinder (mathematical note 3.1, p.47). One end of the cylinder corresponds to infinite distances from the city centre. The other end corresponds to the centre itself which cannot be reached in a finite time.

Radial paths in the field are transformed into the generators of the cylinder, which are straight lines. When this cylinder is slit along one of its generators, it can be unbent into a plane. On this plane, geodesic paths are straight lines. When the cylinder is put back together these straight lines become helices. The geodesics on the cylinder are thus helices, which are the images of the spiral minimum paths in this field mentioned earlier (figure 2.8). In a similar manner we show that the time surface for the velocity field $V(r) = \omega r^p$, where $0 < p < 1$ is a *cone* whose apex is the image point of the city centre (mathematical note 3.1, p.48). By the same argument, we note that geodesic paths on the conic time surface are spiraling space curves. Time on this surface can be evaluated by slitting the surface open along one of its generators, and measuring the Euclidean distance between image points.

Consider now the velocity field $V(r) = ar^2 + b$, where a and b are positive constants. When plotted against distance from the city centre, the cross section of this field takes the form of a parabola. The time surface corresponding to this velocity field is a *sphere* (mathematical note 3.1, pp.49–50). This sphere can be embedded in urban space by multiplying its coordinates by b, the velocity at the city centre. We obtain a sphere in urban space which is tangential to the urban plane at the city centre. The transformation can now be represented by a stereographic projection.

The geodesics on the spherical time surface are great circles. Using a well-known property of stereographic projections, we deduce that the minimum paths on the plane must also be circles (Hilbert and Cohn Vossen, 1952, p.250). The family of minimum paths through a given point is thus a coaxial system of circles passing through another fixed point. The family of isochrones from the original point is another family of coaxial circles, each member of which cuts the members of the first family at right angles. These families of minimum paths and isochrones on the urban plane are presented in figure 3.6 for a point east of the city centre.

A realistic time surface was obtained by constructing the appropriate transformation (mathematical note 3.1, pp.45–47) and applying the transformation to the velocity field for Greater Manchester, 1965. The cross section of this time surface is presented in figure 3.7. At large distances from the centre it approaches a plane. As we approach the congested areas of central Manchester, road speeds diminish and the time surface becomes progressively distorted.

We can conclude, therefore, that it is possible to transform a Euclidean plane with a realistic distribution of transport costs into a surface where these costs are uniformly distributed. This transformation allows us to examine the following two conjectures.

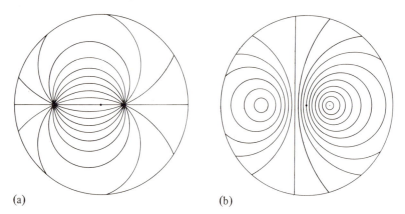

(a)   (b)

Figure 3.6. (a) Minimum paths and (b) isochrones for the velocity field corresponding to a spherical time surface.

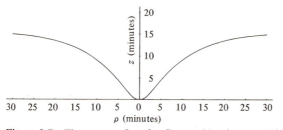

Figure 3.7. The time surface for Greater Manchester, 1965.

3.5 Wardrop's conjecture
A transformation exists which will map a realistic distribution of transport costs on the Euclidean plane into a Euclidean plane with uniform transport facility.

Wardrop (1969) deals with the problem of characterizing curved minimum-cost paths in urban areas by straight line paths on the plane. Wardrop shows that, under certain conditions, a conformal transformation using complex variables will transform minimum-cost curves into straight lines. These conditions require that the function representing the distribution of costs per unit distance in the complex plane will be equal to the absolute value of the derivative of the transformation (Wardrop, 1969, p.186—this book page 155).

Wardrop shows that the only conformal transformations which preserve radial symmetry correspond to velocity fields where the velocity is proportional to some positive power of the distance from the city centre. He then derives a family of minimum paths and isochrones for such a field in the special case where the velocity varies with the square root of the distance from the city centre, $V(r) = \omega r^{1/2}$, where ω is some constant. These are presented in figure 3.8 for a point south of the centre. In the general, nonradially symmetric, case Wardrop suggests that transformations could be found in which speed is reduced as one approaches each of several centres.

None of the examples presented by Wardrop are realistic. In all his cases, velocities can increase without limit. In all realistic velocity fields, however, we must preserve the condition that velocities vary and are bounded from above.

Such a velocity field must have a curved time surface, as depicted in figure 3.7. Clearly, any two time surfaces corresponding to the same velocity field must be related by a length-preserving transformation—the travel time between corresponding pairs of points being equal.

The *Gaussian curvature* of a surface is the product of the two principal curvatures at a point. We note that if two surfaces are related by a length-

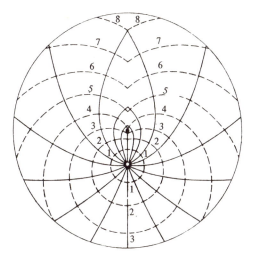

Figure 3.8. Minimum paths and isochrones in the velocity field $V(r) = \omega r^{1/2}$ (from Wardrop, 1969).

preserving transformation, then their Gaussian curvatures at corresponding points must be equal (Hilbert and Cohn Vossen, 1952, p.260). The Euclidean plane has zero Gaussian curvature everywhere, so the only time surfaces that are related to the Euclidean plane by a length-preserving transformation must also have zero Gaussian curvature everywhere.
When the velocity of travel depends only on the distance from a fixed point, the only such surfaces are cylinders and cones. These surfaces correspond to velocity fields where the speed of travel is proportional to some power of the distance from the fixed point. This family is, therefore, the family appropriate to the transformation presented by Wardrop.

The Gaussian curvature of the time surface for Greater Manchester is positive at the point corresponding to the city centre. This arises because the central velocity is strictly positive. The curvature decreases as we move out from the centre and reaches zero at the distance corresponding to the point of inflection in the cross-sectional curve depicted above. The curvature is then negative for the rest of the range but approaches zero at large distances from the centre, where the time surface approaches a plane. Thus the time surface does not have zero Gaussian curvature everywhere and therefore cannot be mapped onto the Euclidean plane by a length-preserving transformation. Thus Wardrop's transformation cannot be applied to velocity fields which satisfy realistic conditions.

3.6 Bunge's conjecture

A transformation exists which will map a realistic density and a realistic distribution of travel costs into a curved surface with uniform density and a uniform distribution of travel costs.

The solution of Tobler's problem discussed earlier is an encouraging sign for Bunge in his search for a transformation which will make the patterns postulated by Christaller, and the other theoretical geographers, visible (Bunge, 1964, pp.21–24):

> "*From around a point,* Tobler has brilliantly, formally and completely solved the problem ... Tobler's work is not a complete solution because we not only need to straighten out the shape of space around a single point but between all points. In other words, around a single point dealing with problems like Thünen's initial statement, a solution has been found; but around many points problems like Christaller's we have not achieved a solution ... Since it has been proved that the traditional geographic map cannot hold the solution to our space straightening problem, what will? It seems to me that the mapping will have to be some object in hyper-space. I have been struck with this notion, unable to advance for three or four years. Tobler does not warm to it so I do not trust it but can offer no alternative".

Clearly, then, Bunge's object in hyperspace must also be a time surface. We first establish a necessary condition for Bunge's conjecture to be true.

To establish this condition, we consider the case where the velocity of travel does not depend on the direction of travel. In this case, the transformation must be locally a scale transformation. It must map a small distance on the Euclidean plane into a small distance on the surface, the scaling factor being the reciprocal of the local velocity of travel. It must also preserve angles between curves. Thus small triangles on the Euclidean plane are mapped into similar triangles on the time surface. The area of the triangle on the time surface must be equal to the area of the original triangle multiplied by the square of the reciprocal of the local velocity of travel (mathematical note 3.1, p.47).

The density of population, at a point on the geographical plane, is the number of people within a unit area around that point. Areas are shrunk by the transformation by the square of the reciprocal of the local velocity. The population density on the time surface must, therefore, be equal to the corresponding density on the original plane multiplied by the square of the velocity of travel. Clearly, then, once the distribution of velocities or costs on the geographical plane is specified, the transformation of densities to the time surface is determined.

Bunge's conjecture requires that the density on the time surface be uniform. For this to be the case, the density on the geographical plane must vary inversely with the square of the velocity, or cost of travel, at any location. This is an unrealistic requirement. To show that this condition is not satisfied we examine the relationship between the velocity field and the density of the residences of commuters in the Greater Manchester urban area for 1965. This density was obtained by fitting a smooth curve to the distribution of residences (see pages 59–60). The comparison is presented in figure 3.9.

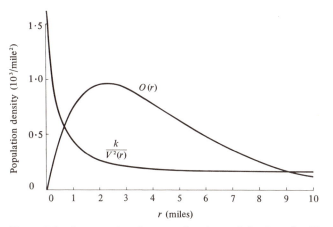

Figure 3.9. A comparison between the observed density of residences of commuters and the density required by Bunge's conjecture in Greater Manchester, 1965.

As can be seen from figure 3.9, the observed density bears little relation to the square of the reciprocal of the local velocity when both are shown as functions of distance from the city centre. At the centre the conjectured density peaks while the observed density is at a minimum. As distance from the centre increases, the conjectured density declines rapidly while the observed density reaches a maximum. At large distances from the centre, the conjectured density reaches a positive constant while the observed density approaches zero. Clearly, then, the required condition is not satisfied and we must conclude that Bunge's conjecture is false.

Close examination of data from studies of the distribution of population for a variety of cities, discussed in the next chapter, reveals that the existence of a maximum density of population at a positive distance from the city centre is a common phenomenon.

3.7 Conclusion

In view of the results of the present chapter it is impossible to retain all three spatial assumptions: the assumption of the Euclidean plane, the assumption of uniform densities, and the assumption of uniform transport facility. In particular the refutation of Wardrop's conjecture precludes the possibility of constructing a flat map of a city which correctly represents travel time. However, since Warntz's conjecture is true we can construct a curved surface which represents travel time. Tobler's transformation enables us to transform a nonuniform distribution on the Euclidean plane into a uniform distribution on the Euclidean plane. This enables us to adapt von Thünen's theory of agricultural production in order to deal with a nonuniform distribution of resources. The most serious implications follow from the refutation of Bunge's conjecture. Since it is impossible to retain both the assumption of uniform densities and the assumption of uniform transport facility even if a curved surface is adopted, we will not be able to use transformations to apply the theories of Lösch and Christaller to realistic environments. So we can never expect to observe the pattern of hexagonal market areas predicted by these theories, however much we try to distort the map. The spatial assumptions of these theories must therefore be relaxed.

If we are not interested in transportation, we can employ geometrical transformations to test geographical theories which assume uniform densities on the Euclidean plane. If we wish to study transportation in a geometrical framework, we must develop non-Euclidean theories of geography. For even with the application of transformations to geographical space, Alonso's comment still applies (quoted in Warntz, 1967, p.9):

"Some wrinkles will not be ironed out".

Mathematical note 3.1. *Time surfaces for radially symmetric velocity fields*

This note describes a procedure for constructing time surfaces from radially symmetric velocity fields. The same procedure also applies to the construction of cost surfaces from cost fields. We first discuss a basic theorem and then illustrate its application in a series of examples.

A *time surface* for a given field is a surface with the property that travel time between two points on any path in the field will be identical with the length of the image of this path between the corresponding points on the time surface.

We are given a polar coordinate system, (r, θ), for the urban plane on which a velocity field, $V(r)$, has been defined; and a cylindrical coordinate system, (ρ, z, ϕ), for the space in which the time surface is located. These coordinate systems are illustrated in figure 3.10.

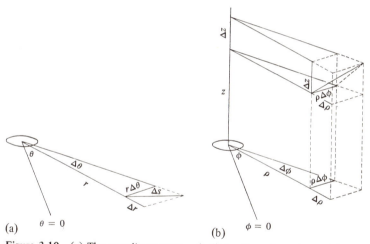

Figure 3.10. (a) The coordinate system (r, θ) for the urban plane. (b) The coordinate system (ρ, z, ϕ) of the space in which the time surface is located.

Theorem 3.1. Let

$$\phi = \theta , \tag{3.1}$$

$$\rho = \frac{r}{V} , \tag{3.2}$$

and

$$z = \int \frac{1}{V^2}\left[2rV\frac{\mathrm{d}V}{\mathrm{d}r} - r^2\left(\frac{\mathrm{d}V}{\mathrm{d}r}\right)^2 \right]^{\frac{1}{2}} \mathrm{d}r + C . \tag{3.3}$$

For any given C these equations define a transformation

$$T_V : (r, \theta) \rightarrow (\rho, z, \phi) .$$

The transformation T_V maps travel time on any path P on the urban plane onto the length of the image of that path, $T_V(P)$.

Proof. It is required to show that

$$\int_P \frac{ds}{V} = \int_{T_V(P)} dt ,$$
(3.4)

where the left side of equation (3.4) denotes travel time on a path P on the urban plane, and the right side denotes the total length of the image of this path under the given transformation.

Given the polar coordinate system on the plane, we can rewrite the left side of equation (3.4) as

$$\int_P \frac{ds}{V} = \int_P \frac{1}{V}\left[1 + r^2\left(\frac{d\theta}{dr}\right)^2\right]^{\frac{1}{2}} dr = \int_P \left[\frac{1}{V^2} + \frac{r^2}{V^2}\left(\frac{d\theta}{dr}\right)^2\right]^{\frac{1}{2}} dr .$$
(3.5)

We have from equation (3.3)

$$\frac{dz}{dr} = \frac{1}{V^2}\left[2rV\frac{dV}{dr} - r^2\left(\frac{dV}{dr}\right)^2\right]^{\frac{1}{2}} ,$$
(3.6)

and

$$\left(\frac{dz}{dr}\right)^2 = \frac{1}{V^4}\left[2rV\frac{dV}{dr} - r^2\left(\frac{dV}{dr}\right)^2\right] .$$
(3.7)

Since $\rho = \dfrac{r}{V}$ by equation (3.2), we have

$$\frac{d\rho}{dr} = \frac{V - r\dfrac{dV}{dr}}{V^2} ,$$
(3.8)

and

$$\left(\frac{d\rho}{dr}\right)^2 = \frac{V^2 - 2rV\dfrac{dV}{dr} + r^2\left(\dfrac{dV}{dr}\right)^2}{V^4} .$$
(3.9)

Thus from equations (3.7) and (3.9) we obtain

$$\left(\frac{dz}{dr}\right)^2 + \left(\frac{d\rho}{dr}\right)^2 = \frac{1}{V^2} .$$
(3.10)

Substituting the above expression, together with equations (3.1) and (3.2), into equation (3.5) we obtain

$$\int_P \frac{ds}{V} = \int_P \left[\left(\frac{dz}{dr}\right)^2 + \left(\frac{d\rho}{dr}\right)^2 + \rho^2\left(\frac{d\phi}{dr}\right)^2\right]^{\frac{1}{2}} dr$$

$$= \int_{T_V(P)} \left[\left(\frac{dz}{d\phi}\right)^2 + \left(\frac{d\rho}{d\phi}\right)^2 + \rho^2\right]^{\frac{1}{2}} d\phi = \int_{T_V(P)} dt .$$
(3.11)

The latter equality is obvious, since the expression under the integral sign of the left side is the element of distance in the cylindrical coordinate system (ρ, z, ϕ).

As r and θ vary over the urban plane, the transformation T_V defines a parametric representation of a surface. This surface will be a time surface by definition.

The transformation T_V is conformal, that is it preserves angles between curves wherever the time surface is locally Euclidean. A small triangle on the plane, with edges Δs_1, Δs_2, and Δs_3, is mapped by the transformation T_V into a small triangle on the time surface with edges Δt_1, Δt_2, and Δt_3. By equation (3.11) we have

$$\frac{\Delta t_1}{\Delta s_1} = \frac{\Delta t_2}{\Delta s_2} = \frac{\Delta t_3}{\Delta s_3} = \frac{1}{V} . \tag{3.12}$$

The two triangles are similar and hence the corresponding angles are equal.

A real time surface exists only as long as the discriminant of equation (3.3) is nonnegative, and $V \neq 0$ for $0 < r < \infty$. This is identical with the requirement that

$$0 \leqslant \frac{dV}{dr} \leqslant \frac{2V}{r} \qquad \text{for} \quad 0 < r < \infty . \tag{3.13}$$

When strict inequalities are satisfied dz/dr has a fixed sign, z is a monotonic function of r, and hence the transformation T_V is one-to-one.

We now derive time surfaces for the various velocity fields discussed earlier in this chapter. Since these examples involve radially symmetric velocity fields, their time surfaces must also be radially symmetric. We can therefore characterize these surfaces by a cross-sectional curve in any (ρ, z) plane.

Corollary 3.1. The time surface of the velocity field $V(r) = \omega r$, where ω is some constant, is a *cylinder*.

Proof. From equation (3.2) we obtain

$$\rho = \frac{1}{\omega} . \tag{3.14}$$

Equation (3.3) takes the forms

$$z = \int \frac{dr}{\omega r} + C = \frac{1}{\omega} \ln\left(\frac{r}{r_0}\right) , \tag{3.15}$$

where r_0 is the radius corresponding to $z = 0$. ρ is independent of z, and the time surface is thus a cylinder of radius $1/\omega$ which extends over the the complete z axis, $z = -\infty$ corresponding to $r = 0$, and $z = +\infty$ corresponding to $r = \infty$.

Corollary 3.2. The time surface of the velocity field $V = \omega r^p$, where ω is some constant and $0 < p < 1$, is a *cone*.

Proof. From equation (3.2) we obtain

$$\rho = \frac{r}{V} = \frac{r}{\omega r^p} = \frac{1}{\omega} r^{1-p} \ . \tag{3.16}$$

By equation (3.3)

$$z = \int \frac{1}{\omega^2 r^{2p}} \left(\frac{2r}{\omega r^p} p \omega r^{p-1} - r^2 p^2 \omega^2 r^{2p-2} \right)^{\frac{1}{2}} dr + C \ . \tag{3.17}$$

Let $C = 0$. We then have

$$z = \frac{r^{1-p}}{\omega(1-p)} (2p - p^2)^{\frac{1}{2}} = \frac{\rho}{1-p} (2p - p^2)^{\frac{1}{2}} \ . \tag{3.18}$$

So z is proportional to ρ, and the time surface is a cone.

Travel time on the velocity field $V(r) = \omega r^p$ can be calculated by measuring distances on the conic time surface. We next derive the expression for travel time between two given points with radial coordinates r_1 and r_2 respectively, and angular difference θ_{12} (the definition of the angular difference was given earlier on p.25).

Corollary 3.3. Travel time between two points in the velocity field $V(r) = \omega r^p$ is given by

$$t(r_1, r_2, \theta_{12}) = \frac{1}{\omega(1-p)} \{ r_1^{2-2p} + r_2^{2-2p} - 2r_1^{1-p} r_2^{1-p} \cos [(1-p)\theta_{12}] \}^{\frac{1}{2}} \ . \tag{3.19}$$

Proof. The time surface for this velocity field is a cone by corollary 3.2. The distance R from the apex to any point on the cone can be evaluated from equation (3.18) to yield

$$R = (\rho^2 + z^2)^{\frac{1}{2}} = \frac{\rho}{1-p} \ . \tag{3.20}$$

We can now open the cone along one of its generators to obtain an area shaped like a piece of pie. The angle of opening of the pie-shaped area is $2\pi(1-p)$ as can be easily verified. Through the transformation of the cone into a pie-shaped area the angle ϕ is transformed into the angle α between two generators where

$$\alpha = \phi(1-p) \ . \tag{3.21}$$

Minimum paths on the pie-shaped area are straight lines since the surface lies on the plane.

The time between two points on this surface, (R_1, α_1) and (R_2, α_2), can be evaluated by the cosine rule. Let α_{12} be the angular difference of the two points. Then

$$t^2(R_1, R_2, \alpha_{12}) = R_1^2 + R_2^2 - 2R_1 R_2 \cos\alpha_{12} . \tag{3.22}$$

Substituting for R_1, R_2 and from equations (3.20), (3.21), and (3.16) we obtain the desired result.

Corollary 3.4. The time surface of the velocity field $V(r) = ar^2 + b$, where a and b are positive constants, is a *sphere* of radius $1/\omega$, where $\omega^2 = 4ab$.

Proof. From equation (3.2) we have

$$\frac{dp}{dr} = \frac{1}{V}\left(1 - \rho\frac{dV}{dr}\right) . \tag{3.23}$$

When $V = ar^2 + b$ and $\omega^2 = 4ab$, we obtain

$$V^2\left(\frac{d\rho}{dr}\right)^2 = (1 - 2ar\rho)^2 = 1 - \rho^2\omega^2 . \tag{3.24}$$

Using equation (3.10) we can write

$$\left(\frac{dz}{dr}\right)^2 + \left(\frac{d\rho}{dr}\right)^2 = \left(\frac{d\rho}{dr}\right)^2 \frac{1}{1 - \rho^2\omega^2} . \tag{3.25}$$

Simplifying this equation we can obtain a differential equation in z and ρ

$$\frac{dz}{d\rho} = \rho\left(\frac{1}{\omega^2} - \rho^2\right)^{-\frac{1}{2}} . \tag{3.26}$$

The solution of this differential equation is

$$(z - C)^2 + \rho^2 = \frac{1}{\omega^2} , \tag{3.27}$$

which is the equation of a sphere of radius $1/\omega$, where C is an arbitrary constant. If we select $C = 1/\omega$, the sphere passes through $\rho = 0$, $z = 0$. This sphere then satisfies

$$\left(z - \frac{1}{\omega}\right)^2 + \rho^2 = \frac{1}{\omega^2} . \tag{3.28}$$

Applying equation (3.2) we obtain

$$\rho = \frac{r}{ar^2 + b} . \tag{3.29}$$

Solving equation (3.28) for z, and substituting the above result for ρ, we obtain

$$z = \frac{2ar^2}{\omega(ar^2 + b)} . \tag{3.30}$$

In this example the construction of minimum paths and isochrones is greatly simplified by the observation that we can represent the transformation T_V

as a stereographic projection. If we multiply the variables (ρ, z) of the
time sphere by the central velocity b, we obtain from equation (3.28)

$$\left(bz - \frac{b}{\omega}\right)^2 + (b\rho)^2 = \left(\frac{b}{\omega}\right)^2 . \tag{3.31}$$

This is the equation of a sphere in urban space of radius b/ω, tangential to
the urban plane at the city centre. The transformation moves corresponding
points along the line which joins them to the north pole. This is illustrated
in the figure 3.11.

To verify that T_V is a stereographic projection we must show that

$$\frac{r}{b\rho} = \frac{2b/\omega}{(2b/\omega) - bz} . \tag{3.32}$$

From equation (3.30) we have

$$\frac{2}{\omega} - z = \frac{2b}{\omega(ar^2 + b)} = \frac{2b}{\omega V} , \tag{3.33}$$

so that both sides of equation (3.32) equal V/b.

The time surface for the Greater Manchester velocity field,

$$V(r) = 24 \cdot 9 - 16 \cdot 9 \exp{-(0 \cdot 56r)} ,$$

cannot be obtained by solving equations (3.2) and (3.3) analytically.
Instead, it is obtained by numerical approximation. From equation (3.2)
we derive ρ as a function of r,

$$\rho = \frac{r}{a - b \exp(-cr)} \tag{3.34}$$

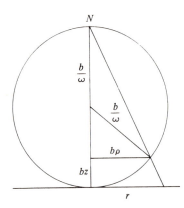

Figure 3.11. Diagram represents a sphere in urban space tangential to the urban plane
at the city centre.

Given the parameters of the velocity field, we can evaluate the integrand of equation (3.3) at any point; z is then obtained as a function of r, $z = z(r)$, by any of the known methods of numerical approximation of integrals. Since the value of the integrand approaches the r axis more rapidly than any power of r, as r approaches infinity the values obtained for z are bounded from above. The time surface thus possesses an asymptotic plane at the supremum of the z values. For each r, the above procedure produces a value for ρ and a corresponding value for z, thus giving us a cross-sectional curve of the time surface. The cross-sectional curve of the 1965 Greater Manchester time surface is presented in figure 3.7.

A continuous model of spatial interaction

4.1 Trip distribution

The methods for calculating travel time or travel cost between locations in geographical space, using a continuous field representation, make it possible to extend many of the concepts and techniques which have been previously restricted to the network representation of space. In particular we can apply many of the mathematical models which have flourished during the past two decades of transportation studies. Until recently these models have remained quite separate from other work in spatial analysis. Much of this modelling effort has been focused on detailed specifications at the expense of obtaining general results and global patterns. By relating the models to the underlying geography, and by applying continuous methods, we move away from detailed data on the one hand and from specific results on the other, into the search for the relationships between basic patterns and the properties of cities and regions.

The aim of this chapter is to develop a continuous model of spatial interaction, a model based on the continuous distribution of phenomena in geographical space, by analogy to the discrete, network-based model derived by Spurkland (1967), Tomlin and Tomlin (1968), and Wilson (1967). This model was chosen for detailed examination and calibrated for the Greater Manchester urban area, 1965. It is a trip-distribution model which uses the density functions describing the distribution of residences and workplaces of car commuters, and the velocity field of a given city, to derive a spatial distribution of trips, accessibilities, and traffic flow[5]. It was seen as particularly important to develop one model in detail and to calibrate it, rather than to illustrate the great variety of discrete models which can be applied in continuous field representations of transport. Shopping models, residential location models, commodity flows, and various other topics in the field of human spatial interaction can be developed along similar lines.

Trip-distribution models have been used extensively in modern transportation studies (see for example Lane *et al.*, 1971). The role of trip-distribution models is to predict the number of trips between different locations in the city—locations being represented by a set of discrete zones. It is assumed that the number of trips originating and terminating in each zone is given. In the case of the journey to work, these numbers can be obtained independently from population characteristics of residential areas, and from employment characteristics of workplaces. It is conventional to assume that the total number of origins and the total number of

[5] The remainder of this chapter is restricted to car travel only, incorporation of other modes must await further developments. The continuous representation of movement may be less suited to the analysis of public rail transit, particularly if the number of links and nodes is relatively small.

destinations are identical in the system under study—all trips originate and terminate within the specified region. Furthermore it is assumed that estimates of the time or cost of travel between all pairs of zones are given. These are estimated by calculating minimum paths between the nodes associated with these zones on the transport network. Wilson further postulates that the average time or money spent on commuting in the region is known. These could be estimated from expenditure patterns of the regional population as a whole.

All this information can be represented as a set of constraints on the possible distribution of trips between the zones. In particular the total number of trips originating in a given zone must sum to the given number of origins there,

$$\sum_{j=1}^{n} T_{ij} = O_i , \qquad i = 1, ..., n ,$$

where T_{ij} is the number of trips between zone i and zone j, and O_i is the given number of origins in zone i.

The total number of trips terminating in a given zone must sum to the given number of destinations in that zone,

$$\sum_{i=1}^{n} T_{ij} = D_j , \qquad j = 1, ..., n ;$$

where D_j is the given number of destinations in zone j. And finally, the total time spent for making all the trips in the system must sum up to the product of the given average time and the total number of trips made[6].

$$\sum_{i=1}^{n} \sum_{j=1}^{n} T_{ij} t_{ij} = T\bar{t} ,$$

where t_{ij} is the travel time between zone i and zone j, $\bar{t}$ the given average travel time in the city, and T the total number of trips.

Given these constraints, there are still many possible distributions $\{T_{ij}\}$ for which they will be satisfied. Any distribution $\{T_{ij}\}$ which satisfies these constraints requires additional information in order to determine where each individual trip originated and terminated, that is to say where each individual lives and works. There exists a unique distribution which requires the maximal amount of additional information to uncover where each individual trip originated and terminated. Such a distribution therefore reflects maximum uncertainty with respect to missing information.

---

[6] This number is also given. It is identical with the total number of origins or destinations in the system, which are obtained by summing the origins or destinations of the individual zones,

$$\sum_{i=1}^{n} O_i = \sum_{j=1}^{n} D_j = T .$$

Let us define

$$p_{ij} = \frac{T_{ij}}{T}$$

as the proportion of trips which originate in zone i and terminate in zone j. Then, if these proportions are known, the amount of information necessary to determine where a single individual lives and works is given by the entropy (see mathematical note 4.1):

$$H = -\sum_{ij} p_{ij} \ln p_{ij} ,$$

the p_{ij} being defined so that they satisfy the condition of summing to unity. We can regard p_{ij} as the probability that a random individual will both live in zone i and work in zone j.

The information necessary to determine where all of the individuals live and work is given by the entropy H multiplied by the total number of trips T, which we have assumed is known.

We thus require that distribution of trips $\{T_{ij}\}$ which maximizes the entropy

$$H = \sum_{i=1}^{n} \sum_{j=1}^{n} p_{ij} \ln p_{ij} ,$$

subject to the three constraints formulated above. This problem can be solved as an ordinary constrained maximisation problem to yield an expression for the distribution of trips:

$$T_{ij} = A_i B_j O_i D_j \exp(-\mu t_{ij}) ,$$

where the A_i's, B_j's, and μ are obtained by solving the $2n+1$ constraint equations. The above expression is a trip-distribution function which predicts the number of trips between every pair of zones from the given information. This type of model has been used in several transportation studies. A survey of the applications of the model is provided by Wilson (1970).

The continuous model exactly parallels the discrete model described above. When urban space is represented as a field we need no longer consider a zoning system or a transport network. The distributions of population and employment are given as density functions, and the travel time or travel cost between each pair of locations is obtained from the urban velocity or cost field. These, together with the average time or cost of commuting in the region, are then used to derive a measure of the density of trips between locations.

Obtaining values for minimum travel times in an arbitrary velocity field involves a generalization of the procedure described earlier (see also mathematical note 2.1) for radially symmetric fields. The following analysis assumes a radially symmetric city, and therefore uses radially symmetric density functions and a radially symmetric velocity field as inputs. (The continuous trip-distribution model is not restricted to radially symmetric cities. The general, non radially symmetric model is presented in mathematical note 4.2.)

Radially symmetric density functions used in the calibration of this model are discussed in section 4.2 below. The results of the model—the accessibility of locations to employment opportunities, the accessibility of workplaces to residences of commuters (section 4.3), and the spatial distribution of traffic flow in the city (section 4.4)—are also obtained as radially symmetric functions of distance from the city centre.

Consider a polar coordinate system (r, θ), where r is the distance from the city centre, and θ is the angle between the radius vector and a fixed axis. An element of area in this coordinate system is $r\,d\theta\,dr$. We define the *origin density function*, $O(r_1)$, as the number of trips originating in a unit area about a point, r_1 miles away from the city centre. In a similar manner we define the *destination density function*, $D(r_2)$, as the number of trips terminating in a unit area about a point, r_2 miles away from the city centre. In this study, one square mile will be taken as the unit of area, but other measures could be adopted. The unit of area must be small enough to ensure that significant variations in density are not glossed over, but large enough to average out local fluctuations.

We now define the *trip density function*, $T(r_1, \theta_1, r_2, \theta_2)$, as the number of trips from a unit area about a point (r_1, θ_1) to a unit area about a point (r_2, θ_2). Thus the trip density is a *joint* density, having marginal densities $O(r_1)$ and $D(r_2)$ (see for example Feller, 1968, chapter 9, for discussion of joint and marginal densities). The dimensions of the trip density function are number per unit area per unit area. If the trip density function is defined for a particular unit of area, then doubling the unit will quadruple the trip density between every pair of points (r_1, θ_1) and (r_2, θ_2). As in the discrete formulation, it is assumed that all trips take place during a specified time period, in this case the morning rush hour. Travel time between any pair of points is computed from a velocity field corresponding to rush-hour conditions and is denoted $t(r_1, \theta_1, r_2, \theta_2)$. The expression for travel time and the expression for trip density are assumed to be invariant under rotations and reflections in the radially symmetric model. Given these definitions we can now write the continuous analogue of Wilson's maximum entropy model. A trip density, $T(r_1, \theta_1, r_2, \theta_2)$, is obtained by maximizing the entropy,

$$H = -\int_0^\infty \int_0^\infty \int_0^{2\pi} \int_0^{2\pi} P(r_1, \theta_1, r_2, \theta_2) \ln P(r_1, \theta_1, r_2, \theta_2) r_1 r_2 \, d\theta_1 \, d\theta_2 \, dr_1 \, dr_2 \ ,$$

where

$$P(r_1, \theta_1, r_2, \theta_2) = \frac{T(r_1, \theta_1, r_2, \theta_2)}{T} \ ,$$

subject to the following constraints:

The total number of trips from a unit area at (r_1, θ_1) to all other areas must be the density of origins there,

$$\int_0^\infty \int_0^{2\pi} T(r_1, \theta_1, r_2, \theta_2) r_2 \, d\theta_2 \, dr_2 = O(r_1) \qquad \text{for all } r_1, \theta_1 \ .$$

Similarly, the total number of trips originating in the entire city and terminating in a unit area at (r_2, θ_2) is the density of destinations there,

$$\int_0^\infty \int_0^{2\pi} T(r_1, \theta_1, r_2, \theta_2) r_1 \, d\theta_1 \, dr_1 = D(r_2) \qquad \text{for all } r_2, \theta_2 \ .$$

And finally, the total expenditure on transport in the city is the product of the average expenditure, $\bar{t}$, and the total number of trips, T, in the system:

$$\int_0^\infty \int_0^\infty \int_0^{2\pi} \int_0^{2\pi} T(r_1, \theta_1, r_2, \theta_2) t(r_1, \theta_1, r_2, \theta_2) r_1 r_2 \, d\theta_1 \, d\theta_2 \, dr_1 \, dr_2 = T\bar{t} \ .$$

The solution to this maximization problem can be obtained using calculus of variations methods. It yields a trip density function of the form[7]

$$T(r_1, \theta_1, r_2, \theta_2) = A(r_1) B(r_2) O(r_1) D(r_2) \exp\left[-\mu t(r_1, \theta_1, r_2, \theta_2)\right] \ ,$$

where $A(r_1)$ and $B(r_2)$ are functions associated with location only. These functions, the balancing factor for origins and the balancing factor for destinations respectively, can be obtained by rewriting the origin and destination constraints in the form

$$\frac{1}{A(r_1)} = \int_0^\infty \int_0^{2\pi} B(r_2) D(r_2) \exp\left[-\mu t(r_1, \theta_1, r_2, \theta_2)\right] r_2 \, d\theta_2 \, dr_2 \ ,$$

$$0 \leqslant r_1 < \infty \ ,$$

and

$$\frac{1}{B(r_2)} = \int_0^\infty \int_0^{2\pi} A(r_1) O(r_1) \exp\left[-\mu t(r_1, \theta_1, r_2, \theta_2)\right] r_1 \, d\theta_1 \, dr_1 \ ,$$

$$0 \leqslant r_2 < \infty \ .$$

The above two equations, together with the constraint on the total expenditure on transport, can now be used to obtain the forms for the balancing factors $A(r_1)$ and $B(r_2)$ (as functions of distance from the city centre) and the value of the parameter μ by using numerical methods for integration and iterative techniques for solving the equations simultaneously. We thus obtain a trip density function which can be evaluated for any pair of locations.

   The trip density function is proportional to a measure of separation between the two locations, $\exp\left[-\mu t(r_1, \theta_1, r_2, \theta_2)\right]$. This means that the number of trips falls off at a negative exponential rate as the time of

[7] The solution of this problem is a special case of the general model described in mathematical note 4.2. The functions $A(r_1)$ and $B(r_2)$ are shown to be independent of the angular coordinates in mathematical note 4.3.

travel between the points increases. When μ is large, trips fall off at a rapid rate and people do not commute very far. When μ is small, trips fall off slowly and people travel greater distances. The value of μ is heavily dependent on the average expenditure on transport in the system, $\bar{t}$. When $\bar{t}$ is small, people do not travel far, and μ is large. When it is large, people travel further, and μ is small.

The trip density function is also related directly to the densities of the origins and the destinations at the end points. When there are more residences at the origin area or more workplaces at the destination area, we can expect more trips to take place between the two areas. The balancing factors $A(r_1)$ and $B(r_2)$ are functions of location only, depending only on the coordinates of the points, and are independent of the unit of area. In other words the balancing factors must be regarded as intensive variables in the sense discussed by Lowry (1968, p.152). The form of the balancing factors and their interpretation is deferred to section 4.3 below.

This model was calibrated for the Greater Manchester urban area and used data and results obtained by the SELNEC (South East Lancashire–North East Cheshire) Transportation Study for 1965. The parameter μ was estimated and the balancing factors were derived as functions of distance from Central Manchester. The computer program for the calibration of the model and for obtaining the results of this and the following chapter will be found in the Appendix (pages 90–107).

In order to test the accuracy of our predictions of the distribution of trips, we have compared it with the results obtained by SELNEC. A useful criterion is to compare the travel time distributions predicted by both models (see Wilson *et al.*, 1969). This comparison is presented in figure 4.1.

As can be observed from the figure, our model predictions follow closely those of SELNEC. Minor deviations can be detected. Our model

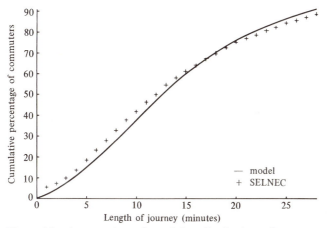

Figure 4.1. A comparison of travel-time distributions of car commuters in the Greater Manchester urban area, 1965.

predicts fewer short commuting trips and more long trips than the SELNEC model. However, despite the simplifying assumptions discussed above, the continuous model yields reasonable results.

4.2 Residential and employment densities of commuters

Having presented the symmetric model for the distribution of trips, we now discuss the radially symmetric density functions for residences and workplaces. We are interested in the distribution of residences in order to overcome the simplifications introduced by the assumption of uniform densities, made by the geographers and regional scientists, which was discussed in the previous chapter. In a similar way we seek to overcome the simplifications of the distribution of employment, made by the urban economists, which we shall discuss in the next chapter. Several authors (see for example Berry *et al.*, 1963; Cassetti, 1967; Stewart and Warntz, 1968; Bussière and Snickars, 1970; and Niedercorn, 1971) describe the distribution of population in a city by a negative exponential function:

$$d(r) = a\exp(-br) \ ,$$

where a is determined by the total population of the city and b is the rate of relative decrease of density with respect to distance from the city centre. These authors all draw upon the detailed empirical work of Colin Clark (Clark, 1951). However, in all the cities studied by Clark the population density is observed to decrease as the city centre is approached. Clark points out that the negative exponential model does not apply in the central area of cities, which is not available for residence. The process of urban decentralization was noted some time ago by Homer Hoyt (Hoyt, 1940, p.270):

"The old compact cities have burst, leaving a vacuum between the main business centre and the outer area of new residential growth".

A recent reappraisal of Clark's work was made by Newling (1969). According to Newling the essential idea is that urban population densities are spatially systematic, the negative exponential formulation being only a good approximation to a more relevant model. In his paper he says

"an alternative hypothesis of the spatial variation of urban population densities is proposed, which views the pattern of density within and beyond the limits of the central business district as a continuum, and which can be placed in a dynamic framework to provide for the emergence of a density crater in the central business district" (Newling, 1969, p.242).

A more recent phenomenon is the tendency for employment to decentralize; a detailed analysis of the occurrence of this effect in England is given by Hall *et al.* (1973).

Similar patterns emerge for distributions of car-using residents in Greater Manchester. We plotted the total number of residences within each circle, centered at the town hall. The distribution of workplaces of car users was plotted in the same way. Both of these distributions, together with the calibrated curves are illustrated in figure 4.2.

In each case we obtained maximum likelihood estimates for the parameters $a, b,$ and c of a gamma distribution, with a density function of the form

$$d(r) = ar^b \exp(-cr) .$$

For residences b is positive, so that the central density is zero. For the distribution of workplaces, b is negative. In this case the central density is infinite, the density of workplaces declining more rapidly than an exponential function.

The resulting density functions are

$$O(r) = 1164r^{0 \cdot 982} \exp(-0 \cdot 439r)$$

for origins or residences of commuters, and

$$D(r) = 4677r^{-0 \cdot 451} \exp(-0 \cdot 298r)$$

for destinations or workplaces of commuters.

In order to illustrate the radially symmetric properties of the density functions, we present them as surfaces in isometric form in figures 4.3 and 4.4.

In order to run the trip-distribution model it was necessary to normalize the distribution of origins and destinations to the same total number. We notice, however, in figure 4.2, that at any radius the cumulative number of workplaces always exceeds the cumulative number of residences.

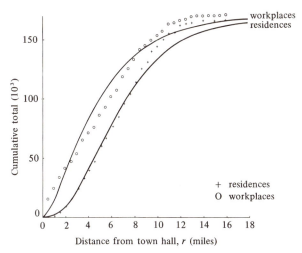

Figure 4.2. The distributions of residences and workplaces of car commuters in Greater Manchester, 1965.

There are more destinations than origins within any circle about the city centre. The cumulative number of workers living within 20 miles of the centre of Manchester and commuting to work by car is found from figure 4.2 to be 166353. The cumulative number of jobs taken by car commuters up to this radius is 168698. We assume that all the workers within the 20 mile contour also work within this contour and taken together, their workplaces are distributed randomly amongst those available. We therefore reduced the density of jobs by the ratio of these two numbers to obtain equal numbers for the total number of jobs and the total number of workers within the 20 mile contour.

The development of the general continuous model for the distribution of trips involves the relaxation of the radial symmetry assumption for population densities. Dacey (1968), for example, considers a city composed of several centres of population. He then assumes that the whole population

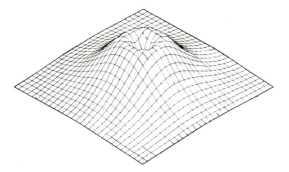

Figure 4.3. The density of residences of car commuters in Manchester, 1965.

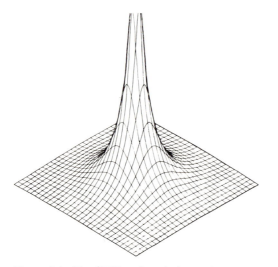

Figure 4.4. The density of workplaces of car commuters in Manchester, 1965.

can be divided into several classes, each class being distributed independently around one of the centres. The density function for the whole population is then obtained by adding together the density functions corresponding to each centre. This kind of analysis has an obvious application for modelling a regional distribution of population. It might be possible to associate centres of employment with different towns and decompose the distribution of residences by means of their employment location. Gurevich and Saushkin (1966) follow a different approach. They consider a single-centre city where the population density declines steadily away from this centre, and discuss five types of cities, each representing a different pattern of radial and angular variation in the population density. These density patterns may be relevant to describing the distribution of population in large conurbations, and more appropriate for a general, nonradially symmetric model. It is also possible to consider densities which are not described by a particular functional form by using contouring procedures to transform discrete field data into a continuous form (see for example Rhind, 1971).

4.3 Balancing factors and accessibilities

This section presents the results of calibrating the radially symmetric trip-distribution model for Greater Manchester and interprets these results using basic concepts of spatial interaction. We present the derived forms for the balancing factors, $A(r_1)$ and $B(r_2)$, and study their behaviour. While balancing factors form an integral part of the trip-distribution model as they are required in order to obtain the density of trips between locations, they also function as measures of the accessibilities of locations. The reciprocals of the balancing factors are interpreted as measures of accessibility. This interpretation differs from simpler measures of accessibility, which usually depend on either the distribution of workplaces or the distribution of residences. In this model, the accessibility to jobs depends both on the distribution of jobs and on the distribution of workers competing for these jobs. Similarly the accessibility of workplaces to residences depends on the distribution of residences, and the distribution of workplaces competing for their services.

Recall the equations for the balancing factors in the radially symmetric model derived in section 4.1. For illustrative purposes we rewrite these equations as

$$\frac{1}{A(r_1)} = \int_0^\infty B(r_2)D(r_2)r_2 \int_0^{2\pi} \exp\left[-\mu t(r_1,\theta_1,r_2,\theta_2)\right] d\theta_2\, dr_2 \ ,$$

$$0 < r_1 < \infty \ ,$$

$$\frac{1}{B(r_2)} = \int_0^\infty A(r_1)O(r_1)r_1 \int_0^{2\pi} \exp\left[-\mu t(r_1,\theta_1,r_2,\theta_2)\right] d\theta_1\, dr_1 \ ,$$

$$0 < r_2 < \infty \ .$$

The expression,

$$\int_0^{2\pi} \exp[-\mu t(r_1, \theta_1, r_2, \theta_2)] \, d\theta_2 \; ,$$

can be interpreted as the proximity of r_1 and r_2. It will be large when $|r_1 - r_2|$ is small. The left-hand term of the expression for $1/A(r_1)$ increases with the number of job opportunities available at r_2. Therefore $1/A(r_1)$ will be large when there are many job opportunities in the proximity of r_1. By a similar argument $1/B(r_2)$ will be large when there are many residences of workers in proximity to r_2. $1/B(r_2)$ is therefore a measure of the competition of all workers for jobs at r_2. $1/A(r_1)$, therefore, has the natural interpretation as the *relative accessibility to job opportunities* at a location, r_1 miles from the city centre. It is directly related to the number of jobs in close proximity to r_1, and is inversely related to the number of workers living in close proximity to these jobs. By a similar argument $1/B(r_2)$ has the interpretation of the *relative accessibility to workers* of a location, r_2 miles from the city centre. It is directly related to the number of workers living in close proximity to r_2, and is inversely related to the number of jobs in close proximity to these workers.

$1/A(r_1)$ and $1/B(r_2)$ are interpreted as relative values since they are both determined only up to a constant multiple, say k. Thus the balancing factors $kA(r_1)$ and $(1/k)B(r_2)$ will yield the same distribution of trips. To obtain values for these functions, an initial value, say the value of $B(O)$, must be fixed arbitrarily. A form of the discrete trip-distribution model which does not contain an arbitrary scaling factor was derived by Kirby, 1970 (a summary discussion of Kirby's modification of the discrete model and the continuous analogue of Kirby's model appear in mathematical note 4.4). Kirby defines the normalizing factors for the discrete model. These turn out to be constant multiples of the balancing factors. They appear when the Wilson model discussed earlier is written in a slightly different form. Since these normalizing factors do not contain a scaling factor, their reciprocals can be referred to as *accessibilities* rather than relative accessibilities. The accessibilities to jobs and residences for the Greater Manchester urban area were obtained from the continuous model and are presented in figure 4.5.

The surprising feature of these results is that the maximum accessibility to jobs is not at the centre of the city but in a ring approximately a mile away from the centre, in spite of the large number of jobs available at the centre. This is due to the fact that congestion in the central area reduces the proximity of very central locations to jobs outside the centre. The inaccessibility of the centre to jobs for car commuters manifests itself clearly in the existing distribution of residences (figure 4.3) since there are hardly any residences of car commuters in close proximity to the centre.

The reader should also note that the accessibility of jobs to residences, peaks quite a distance away from the centre. This is due in large measure to the concentration of residences in this ring. It indicates that those industries and businesses seeking to maximize access to residences would benefit by moving away from the centre to suburban locations. By comparing the existing density distribution of jobs in figure 4.4 with the distribution of accessibility to residences in figure 4.5, we note that the current density distribution of jobs fails to correspond to the distribution of accessibility. Most jobs are concentrated closer to the centre than to the location most accessible to workers' residences.

It is appropriate at this stage to compare the forms for the balancing factors as they appear in the continuous model, with the corresponding terms in the discrete model. It might be expected that, as we reduce the size of zones used in the discrete model, the values of the balancing factors would approach the values pertaining to the continuous model. This, however, is not the case, the discrete balancing factors implicitly incorporate a measure of zonal area. As the zone size in the discrete model is reduced, the balancing factors will reduce in proportion. In other words, in the discrete model the measures of relative accessibility appear to depend on the zoning system adopted. If we define relative accessibility as the reciprocal of the balancing factor occuring in the continuous model, the corresponding measure in the discrete model is not the reciprocal of the discrete balancing factor. The correct correspondence is obtained by multiplying the reciprocal of the balancing factor by the area of the appropriate zone. The necessity for this arises because in the discrete model the constraints are expressed in terms of zonal totals, whilst in the continuous case, densities are used.

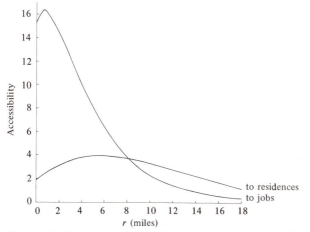

Figure 4.5. The accessibilities of workers to job opportunities, and the accessibility of workplaces to residences in Greater Manchester, 1965.

4.4 The spatial distribution of traffic

A second outcome of the continuous trip-distribution model is the spatial
distribution of traffic in the city. The problem of determining the flow of
traffic parallels the traffic assignment problem encountered in transportation
studies when one uses a network formulation. The problem of traffic
assignment is to determine the total flow of traffic on each link in the
road system. In the discrete formulation the trips between each origin–
destination pair are assigned to the minimum-cost path, and the resulting
loads on each link can then be aggregated. In the continuous formulation
we do not have a network model of the road system. We do have the
densities of trips between locations and the minimum-time paths between
them. The problem of traffic assignment in the continuous formulation is
to aggregate the flow densities along each minimum path passing through a
specified location. This problem has been formulated by Lam and Newell
(1967), who assume that the travel takes place only in radial and
circumferential directions, and derive a continuous approximation for the
flow of traffic in each of these directions. Lam and Newell do not derive
a distribution of trips, assuming that this distribution has been specified.

In the present context we can use the output of our distribution model
to produce the forms for these flows. In the following discussion we
consider only the radial component of flow, $C(r)$, which is the number of
trips crossing a circular cordon r miles from the centre. The function $C(r)$
is referred to as the *cordon-crossing function*. It is composed of three
types of terms: *outgoing trips*, *ingoing trips*, and *double-crossing trips*.

$$C(r) = C_O(r) + C_I(r) + 2C_D(r) ,$$

where $C_O(r)$ is the total number of outgoing trips crossing the cordon in
an outward direction only, $C_I(r)$ the number of ingoing trips crossing the
cordon in an inward direction, and $C_D(r)$ the number of double-crossing
trips, which cross the cordon twice, first in an inward direction and later
in an outward direction. It has been shown (mathematical note 2.1, p.23)
that the nature of minimum paths in the types of velocity fields usually
encountered preclude the possibility of trips which cross a circle originally
in an outward direction and later cross the same circle in an inward
direction. These kinds of trips are therefore not included in the calculation
of the cordon-crossing function.

The functions $C_O(r)$, $C_I(r)$, and $C_D(r)$ can be computed from the trip
density function by integrating it over the appropriate regions. To obtain
the number of outgoing trips, $C_O(r)$, crossing a circle of radius r, we
compute the total number of trips from origins up to radius r to destinations
that lie beyond r. To obtain the ingoing trips, $C_I(r)$, crossing a circle of
radius r, we compute the number of trips between origins that lie beyond
r and destinations that lie inside r. To obtain the number of double-
crossing trips, $C_D(r)$, at radius r, we compute the number of trips between
origins and destinations lying outside r which are separated by an angle

wide enough to allow the minimum-time paths between the two radii to cross the circle of radius r twice. (These integrals are presented in mathematical note 4.4, pp.72–73.)

The forms for the crossing functions derived from the radially symmetric model for Greater Manchester, 1965, are presented in figure 4.6.

It is important to note two features in figure 4.6. First, it can be observed that the number of cordon crossings is influenced mostly by ingoing trips, and therefore peaks where these crossings are at a maximum, that is at about 4 miles from the centre of Manchester. The contribution of outgoing trips is minimal throughout. The contribution of double-crossing trips is only apparent close to the centre, where they reach a value of approximately 30 000 crossings during the morning peak at one mile from the centre. In other words, 15 000 vehicles daily enter a circle of a mile radius and exit it before they reach their destinations.

Second, note that the curves for $2C_D(r)$ and $C(r)$ are shown in broken lines inside a mile radius. This is due to the form of the computer output which specifies crossings of radii at one mile intervals. We therefore extrapolate these curves towards the centre, our only condition being that the outgoing, ingoing, and double-crossing trips will sum up to the cordon-crossing function throughout the range. Consequently we cannot be entirely sure of the behaviour of the cordon-crossing function, $C(r)$, at the city centre. More precise calculations might indicate that these extrapolations need to be revised.

These crossing functions describe the equilibrium distribution of traffic in the city. Their relationships to the flow of traffic on the roads at each location, and consequently to the velocity there, have been studied by Williams and Ortuzar-Salas (1974) (see also Williams, 1974a; 1974b). When these relationships are made operational as capacity-restraint algorithms the continuous model can be used to explore the consequences of large-scale changes in the road system.

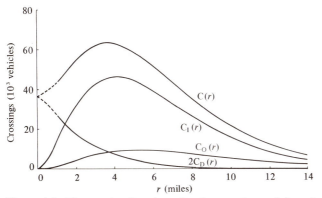

Figure 4.6. The crossing functions, for the morning peak-hour journey to work, for Greater Manchester, 1965.

In this chapter we obtained a continuous form of a spatial interaction model for an urban area. This model enlarges considerably the possibilities for investigating problems of spatial interaction using a geometrical spatial framework. Many interesting problems which were previously hindered by a too simplistic measure of spatial separation, such as airline distance, or by an arbitrary division of geographical space into zones, can now be investigated. The following chapter discusses the application of this model to economic theories concerned with the spatial structure of urban regions.

Mathematical note 4.1. *The derivation of a continuous trip density function with maximum entropy*
In this note we shall obtain a form for the continuous trip density function which satisfies the origin and destinations constraints and the average travel-time constraint, which is of maximum entropy. This trip density function thus uses all the available information given in the form of the constraints, and is maximally noncommital with regard to missing information. This follows from the axiomatic definition of entropy as a unique measure of missing information (Shannon and Weaver, 1964, pp.48–50). We establish the following theorem:

Theorem 4.1. We are given a density function for the origins of trips, $O(x_1, y_1)$, and a density function for the destinations of trips, $D(x_2, y_2)$, over the Euclidean plane, and an average travel time, $\bar{t}$, in some urban region, such that the following constraints are satisfied:

$$\int_0^\infty \int_0^\infty T(x_1, y_1, x_2, y_2) \, dx_2 \, dy_2 = O(x_1, y_1) , \qquad 0 < x_1 < \infty ,$$
$$0 < y_1 < \infty , \qquad (4.1)$$

$$\int_0^\infty \int_0^\infty T(x_1, y_1, x_2, y_2) \, dx_1 \, dy_1 = D(x_2, y_2) , \qquad 0 < x_2 < \infty ,$$
$$0 < y_2 < \infty , \qquad (4.2)$$

and

$$\int_0^\infty \int_0^\infty \int_0^\infty \int_0^\infty T(x_1, y_1, x_2, y_2) t(x_1, y_1, x_2, y_2) \, dx_1 \, dx_2 \, dy_1 \, dy_2 = \bar{t}T , \quad (4.3)$$

where T is the total number of trips in the region. The density of trips from a location around a point (x_1, y_1) to another location about a point (x_2, y_2), which has maximum entropy, takes the form

$$T(x_1, y_1, x_2, y_2) = A(x_1, y_1) B(x_2, y_2) O(x_1, y_1) D(x_2, y_2)$$

$$\exp\left[-\mu t(x_1, y_1, x_2, y_2)\right] \qquad (4.4)$$

where the balancing factors, $A(x_1, y_1)$ and $B(x_2, y_2)$, and the parameter μ are determined by equations (4.1), (4.2), and (4.3). [The balancing factors can be obtained as functions of location by simultaneously solving equations (4.1), (4.2), (4.3), and (4.4).] The solution requires the application of numerical techniques and iterative procedures. The reader will find these described in the Appendix (pp.90–107).

Proof. Define the normalized density function

$$P(x_1,y_1,x_2,y_2) = \frac{T(x_1,y_1,x_2,y_2)}{T} \ . \tag{4.5}$$

The entropy of the density function $P(x_1,y_1,x_2,y_2)$ is given by

$$H = -\int_0^\infty \int_0^\infty \int_0^\infty \int_0^\infty P(x_1,y_1,x_2,y_2)\ln P(x_1,y_1,x_2,y_2)\,dx_1\,dx_2\,dy_1\,dy_2 \ . \tag{4.6}$$

Consider another density function:

$$P^*(x_1,y_1,x_2,y_2) = P(x_1,y_1,x_2,y_2) + \eta(x_1,y_1,x_2,y_2) \ , \tag{4.7}$$

which has the property that

$$T^*(x_1,y_1,x_2,y_2) = TP^*(x_1,y_1,x_2,y_2) \ , \tag{4.8}$$

and satisfies the constraints in equations (4.1), (4.2), and (4.3). The term $\eta(r_1,\theta_1,r_2,\theta_2)$ must be small, in the sense that

$$\left| \frac{\eta(x_1,y_1,x_2,y_2)}{P(x_1,y_1,x_2,y_2)} \right| < \epsilon \ , \tag{4.9}$$

where ϵ is a positive number that is small compared to unity.

Since $T^*(x_1,y_1,x_2,y_2)$ satisfies equations (4.1), (4.2), and (4.3) we deduce that

$$\int_0^\infty \int_0^\infty \eta(x_1,y_1,x_2,y_2)\,dx_2\,dy_2 = 0 \ , \tag{4.10}$$

$$\int_0^\infty \int_0^\infty \eta(x_1,y_1,x_2,y_2)\,dx_1\,dy_1 = 0 \ , \tag{4.11}$$

and

$$\int_0^\infty \int_0^\infty \int_0^\infty \int_0^\infty t(x_1,y_1,x_2,y_2)\eta(x_1,y_1,x_2,y_2)\,dx_1\,dx_2\,dy_1\,dy_2 = 0 \ . \tag{4.12}$$

The entropy of $P^*(x_1,y_1,x_2,y_2)$ is given by

$$H^* = -\int_0^\infty \int_0^\infty \int_0^\infty \int_0^\infty P^*(x_1,y_1,x_2,y_2)\ln P^*(x_1,y_1,x_2,y_2)\,dx_1\,dx_2\,dy_1\,dy_2 \ . \tag{4.13}$$

We show that if $P(r_1,\theta_1,r_2,\theta_2)$ is given by equations (4.4) and (4.5) then, for ϵ sufficiently small, the entropy of $P(x_1,y_1,x_2,y_2)$ exceeds that of $P^*(x_1,y_1,x_2,y_2)$:

$$H \geqslant H^* \ . \tag{4.14}$$

Let us denote $P(x_1,y_1,x_2,y_2)$ by P_{12}, $P^*(x_1,y_1,x_2,y_2)$ by P_{12}^*, and $\eta(x_1,y_1,x_2,y_2)$ by η_{12}. We define a function $F(x_1,y_1,x_2,y_2)$, denoted by F_{12}, such that

$$F_{12} = P_{12}^* \ln P_{12}^* - P_{12} \ln P_{12} \ . \tag{4.15}$$

From equations (4.7) and (4.15) we have

$$\begin{aligned} F_{12} &= (P_{12}+\eta_{12})\ln(P_{12}+\eta_{12}) - P_{12}\ln P_{12} \\ &= P_{12}\ln\left(\frac{P_{12}+\eta_{12}}{P_{12}}\right) + \eta_{12}\ln\left(\frac{P_{12}+\eta_{12}}{P_{12}}\right) + \eta_{12}\ln P_{12} \\ &= P_{12}\ln\left(1+\frac{\eta_{12}}{P_{12}}\right) + \eta_{12}\ln\left(1+\frac{\eta_{12}}{P_{12}}\right) + \eta_{12}\ln P_{12} \ . \end{aligned} \tag{4.16}$$

We expand the logarithmic expression in equation (4.16) in powers of η_{12}/P_{12}. For ϵ sufficiently small we need only retain terms of the first and second degree. Equation (4.16) thus reduces to

$$F_{12} = \eta_{12} + \eta_{12}\ln P_{12} + \frac{\eta_{12}^2}{2P_{12}} + O(\epsilon^3) \ , \tag{4.17}$$

where $O(\epsilon^3)$ is a term less than ϵ^3 by equation (4.9). From the definition of the function $F(x_1,y_1,x_2,y_2)$ in equation (4.15) we observe

$$H-H^* = \int_0^\infty \int_0^\infty \int_0^\infty \int_0^\infty F(x_1,y_1,x_2,y_2)\,dx_1\,dx_2\,dy_1\,dy_2 \ . \tag{4.18}$$

Using equation (4.17) we obtain

$$\begin{aligned} H-H^* &= \int_0^\infty \int_0^\infty \int_0^\infty \int_0^\infty \eta(x_1,y_1,x_2,y_2)\,dx_1\,dx_2\,dy_1\,dy_2 \\ &+ \int_0^\infty \int_0^\infty \int_0^\infty \int_0^\infty \eta(x_1,y_1,x_2,y_2)\ln P(x_1,y_1,x_2,y_2)\,dx_1\,dx_2\,dy_1\,dy_2 \\ &+ \int_0^\infty \int_0^\infty \int_0^\infty \int_0^\infty \left[\frac{\eta^2(x_1,y_1,x_2,y_2)}{2P(x_1,y_1,x_2,y_2)} + O(\epsilon^3)\right] dx_1\,dx_2\,dy_1\,dy_2 \ . \end{aligned} \tag{4.19}$$

We show that the first two terms in the right-hand side of equation disappear.

From equation (4.10) we deduce

$$\int_0^\infty \int_0^\infty \int_0^\infty \int_0^\infty \eta(x_1,y_1,x_2,y_2)\,dx_1\,dx_2\,dy_1\,dy_2 = 0 \ . \tag{4.20}$$

From equation (4.4) and (4.5) we have

$$\int_0^\infty \int_0^\infty \int_0^\infty \int_0^\infty \eta(x_1,y_1,x_2,y_2)\ln P(x_1,y_1,x_2,y_2)\,dx_1\,dx_2\,dy_1\,dy_2$$

$$= \int_0^\infty \int_0^\infty \int_0^\infty \int_0^\infty \left[\eta(x_1,y_1,x_2,y_2)\ln\{A(x_1,y_1)O(x_1,y_1)\}\right.$$

$$+\eta(x_1,y_1,x_2,y_2)\ln\{B(x_2,y_2)D(x_2,y_2)\}$$

$$\left.-\eta(x_1,y_1,x_2,y_2)\mu t(x_1,y_1,x_2,y_2)-\eta(x_1,y_1,x_2,y_2)\ln T\right]dx_1\,dx_2\,dy_1\,dy_2$$

$$= \int_0^\infty \int_0^\infty \ln[A(x_1,y_1)O(x_1,y_1)]\left[\int_0^\infty \int_0^\infty \eta(x_1,y_1,x_2,y_2)\,dx_2\,dy_2\right]dx_1\,dy_1$$

$$+\int_0^\infty \int_0^\infty \ln[B(x_2,y_2)D(x_2,y_2)]$$

$$\times\left[\int_0^\infty \int_0^\infty \eta(x_1,y_1,x_2,y_2)\,dx_1\,dy_1\right]dx_2\,dy_2$$

$$-\mu\int_0^\infty \int_0^\infty \int_0^\infty \int_0^\infty \eta(x_1,y_1,x_2,y_2)t(x_1,y_1,x_2,y_2)\,dx_1\,dx_2\,dy_1\,dy_2$$

$$-\ln T\int_0^\infty \int_0^\infty \int_0^\infty \int_0^\infty \eta(x_1,y_1,x_2,y_2)\,dx_1\,dx_2\,dy_1\,dy_2$$

$$= 0 \tag{4.21}$$

by equations (4.10), (4.11), (4.12), and (4.20).

So, in view of equations (4.20) and (4.21), equation (4.19) reduces to

$$H-H^* = \int_0^\infty \int_0^\infty \int_0^\infty \int_0^\infty \left[\frac{\eta^2(x_1,y_1,x_2,y_2)}{2P(x_1,y_1,x_2,y_2)}+O(\epsilon^3)\right]dx_1\,dx_2\,dy_1\,dy_2 . \tag{4.22}$$

For ϵ sufficiently small

$$\frac{\eta^2(x_1,y_1,x_2,y_2)}{2P(x_1,y_1,x_2,y_2)}+O(\epsilon^3) \geqslant 0 \tag{4.23}$$

throughout the whole range of integration of equation (4.22). Hence we deduce that $H \geqslant H^*$, as required.

This trip density function does not rely on radial symmetry. When the origin and destination density functions, and the velocity field in a given region, are given as radially symmetric functions it is preferable to use a polar coordinate system, (r,θ), rather than a cartesian one. An element of area in the polar coordinate system is $r\,dr\,d\theta$. The range of integration of r is $0 \leqslant r < \infty$, and the range of integration of θ is $0 \leqslant \theta \leqslant 2\pi$.

In the radially symmetric case, the constraint equations take the form

$$\int_0^\infty \int_0^{2\pi} T(r_1,\theta_1,r_2,\theta_2) r_2 \, d\theta_2 \, dr_2 = O(r_1) , \qquad 0 \leqslant r_1 < \infty , \qquad (4.24)$$

$$\int_0^\infty \int_0^{2\pi} T(r_1,\theta_1,r_2,\theta_2) r_1 \, d\theta_1 \, dr_1 = D(r_2) , \qquad 0 \leqslant r_2 < \infty , \qquad (4.25)$$

and

$$\int_0^\infty \int_0^\infty \int_0^{2\pi} \int_0^{2\pi} T(r_1,\theta_1,r_2,\theta_2) t(r_1,\theta_1,r_2,\theta_2) r_1 r_2 \, d\theta_1 \, d\theta_2 \, dr_1 \, dr_2 = \bar{t} T .$$
$$(4.26)$$

The trip density function takes the form

$$T(r_1,\theta_1,r_2,\theta_2) = A(r_1,\theta_1) B(r_2,\theta_2) O(r_1) D(r_2) \exp[-\mu t(r_1,\theta_1,r_2,\theta_2)] .$$
$$(4.27)$$

Where $T(r_1,\theta_1,r_2,\theta_2)$ and $t(r_1,\theta_1,r_2,\theta_2)$ are given in the context of the radially symmetric model, it is possible to show that the balancing factors depends only on distance from the centre. This result is obtained in mathematical note 4.2.

Mathematical note 4.2. *Balancing factors for the radially symmetric trip-distribution model*
In this note we show that the balancing factors in the radially symmetric model are functions of distance from the city centre and are independent of the angular coordinate. This results simplifies the trip density function and the consequent calculation of the balancing factors.

Theorem 4.2. When the origin and destination density functions are given as radially symmetric functions, $O(r_1)$ and $D(r_2)$ respectively, and the travel time $t(r_1,\theta_1,r_2,\theta_2)$ and the trip density function $T(r_1,\theta_1,r_2,\theta_2)$ are invariant under rotations and reflections, then there exist function $A(r_1)$ and $B(r_1)$ such that the trip density function takes the form

$$T(r_1,\theta_1,r_2,\theta_2) = A(r_1) B(r_2) O(r_1) D(r_2) \exp[-\mu t(r_1,\theta_1,r_2,\theta_2)] . \quad (4.28)$$

Proof. We rewrite equations (4.27):

$$T(r_1,\theta_1,r_2,\theta_2) = A(r_1,\theta_1) B(r_2,\theta_2) O(r_1) D(r_2) \exp[-\mu t(r_1,\theta_1,r_2,\theta_2)] .$$
$$(4.29)$$

For radial travel where $\theta_1 = \theta_2 = \theta$, since $T(r_1,\theta_1,r_2,\theta_2)$ and $t(r_1,\theta_1,r_2,\theta_2)$ are invariant under rotation, we obtain from equation (4.29)

$$A(r_1,\theta) B(r_2,\theta) = A(r_1,0) B(r_2,0) \qquad (4.30)$$

for all angles θ.

Again, since $T(r_1, \theta_1, r_2, \theta_2)$ and $t(r_1, \theta_1, r_2, \theta_2)$ are invariant under rotation and reflection, we can write

$$T(r_1, \theta_1, r_2, \theta_2) = T(r_1, \theta_2, r_2, \theta_1) , \qquad (4.31)$$

and

$$t(r_1, \theta_1, r_2, \theta_2) = t(r_1, \theta_2, r_2, \theta_1) . \qquad (4.32)$$

By equations (4.29), (4.31), and (4.32) we can write

$$A(r_1, \theta_1) B(r_2, \theta_2) = A(r_1, \theta_2) B(r_2, \theta_1) . \qquad (4.33)$$

Since the above equation holds for any pair of angles θ_1 and θ_2 we can also write

$$A(r_1, \theta) B(r_2, 0) = A(r_1, 0) B(r_2, \theta) . \qquad (4.34)$$

From equations (4.30) and (4.34) we deduce

$$A(r_1, \theta) = A(r_1, 0) . \qquad (4.35)$$

We define

$$A(r_1) = A(r_1, 0) \qquad (4.36)$$

and

$$B(r_2) = B(r_2, 0) . \qquad (4.37)$$

So from equations (4.35) and (4.36)

$$A(r_1, \theta) = A(r_1) , \qquad \text{for all } \theta ; \qquad (4.38)$$

and from equations (4.34), (4.35), and (4.37) we obtain

$$B(r_2, \theta) = B(r_2) , \qquad \text{for all } \theta . \qquad (4.39)$$

Mathematical note 4.3. *Normalizing factors for the continuous trip-distribution model*

The normalizing factors for the discrete trip-distribution model were defined by Kirby (1970). They are multiples of the usual balancing factors, A_i and B_j, for the discrete model, but do not contain an arbitrary scaling factor. These normalising factors appear when the model is written in the form

$$T_{ij} = \frac{A_i^* B_j^*}{\gamma T} O_i D_j \exp{-\mu t_{ij}} . \qquad (4.73)$$

The values for the normalizing factors A_i^* and B_j^* are defined by the condition

$$\sum_i A_i^* O_i = \sum_j B_j^* D_j = \gamma T . \qquad (4.74)$$

In view of the origin and destination constraints

$$\sum_i T_{ij} = O_i , \qquad i = 1, ..., n \tag{4.75}$$

and

$$\sum_i T_{ij} = D_j , \qquad j = 1, ..., n \tag{4.76}$$

we can also write the normalizing factors in the form

$$A_i^* = \frac{\sum_j T_{ij}/\exp{-\mu t_{ij}}}{\sum_j T_{ij}} = A_i \sum_j B_j D_j , \tag{4.77}$$

$$B_j^* = \frac{\sum_i T_{ij}/\exp{-\mu t_{ij}}}{\sum_i T_{ij}} = B_j \sum_i A_i O_i , \tag{4.78}$$

and

$$\gamma = \frac{\sum_i \sum_j T_{ij}/\exp{-\mu t_{ij}}}{\sum_i \sum_j T_{ij}} = \frac{\sum_i \sum_j A_i B_j O_i D_j}{T} \tag{4.79}$$

The definition of the normalizing factors $A^*(r_1)$, $B^*(r_2)$, and γ for the continuous trip-distribution model exactly parallels that of Kirby. The continuous versions of equations (4.77), (4.78), and (4.79) produce expressions for γ, $A^*(r_1)$, $B^*(r_2)$. In the radially symmetric case we have

$$\gamma = \frac{4\pi^2}{T} \int_0^\infty \int_0^\infty A(r_1)B(r_2)O(r_1)D(r_2)r_1 r_2 \, dr_1 \, dr_2 , \tag{4.80}$$

$$A^*(r_1) = 2\pi A(r_1) \int_0^\infty B(r_2)D(r_2)r_2 \, dr_2 , \tag{4.81}$$

$$B^*(r_2) = 2\pi B(r_2) \int_0^\infty A(r_1)O(r_1)r_1 \, dr_1 . \tag{4.82}$$

The reciprocals of these normalizing factors, representing the accessibility of residences to job opportunities, and the accessibility of workplaces to workers respectively, are presented in figure 4.5, p.63.

Mathematical note 4.4. *Cordon-crossing functions in a radially symmetric city*

In this note we obtain expressions giving the number of trips crossing a circular cordon about the city centre for the radially symmetric trip-distribution model.

The total number, $C(r)$, of trips crossing a circle of radius r is referred to as the *cordon-crossing function*. The cordon-crossing function is composed of three types of terms

$$C(r) = C_O(r) + C_I(r) + 2C_D(r) , \tag{4.83}$$

where $C_O(r)$ is the total number of trips crossing the circle in an outward direction only, $C_I(r)$ is the number crossing only in an inward direction only, and $C_D(r)$ is the number which cross the circle twice, first in an inward direction and later in an outward direction. It has been shown earlier (mathematical note 2.1, p.23) that the nature of minimum paths in the types of velocity fields usually encountered precludes the possibility of trips which cross a circle originally in an outward direction and later cross the same circle in an inward direction. The functions $C_O(r)$, $C_I(r)$, and $C_D(r)$ can be computed from the trip density function by means of the following integrals:

$$C_O(r) = \int_r^\infty \int_0^r \int_0^{2\pi} \int_0^{2\pi} A(r_1)O(r_1)B(r_2)D(r_2)\exp[-\mu t(r_1,\theta_1,r_2,\theta_2)]$$

$$\times r_1 r_2 \, d\theta_1 \, d\theta_2 \, dr_1 \, dr_2 , \qquad (4.84)$$

$$C_I(r) = \int_0^r \int_r^\infty \int_0^{2\pi} \int_0^{2\pi} A(r_1)O(r_1)B(r_2)D(r_2)\exp[-\mu t(r_1,\theta_1,r_2,\theta_2)]$$

$$\times r_1 r_2 \, d\theta_1 \, d\theta_2 \, dr_1 \, dr_2 , \qquad (4.85)$$

$$C_D(r) = \int_r^\infty \int_r^\infty \int_0^{2\pi} \int_0^{2\pi} A(r_1)O(r_1)B(r_2)D(r_2)\exp[-\mu t(r_1,\theta_1,r_2,\theta_2)]$$
$$\scriptstyle |\bar\theta_1 - \theta_2| > \bar\theta$$

$$\times r_1 r_2 \, d\theta_1 \, d\theta_2 \, dr_1 \, dr_2 , \qquad (4.86)$$

where $\bar\theta$ represents $\bar\theta(r_1,r) + \bar\theta(r_2,r)$.

The expression $|\theta_1 - \theta_2|$ in equation (4.86) denotes the absolute value of the difference between θ_1 and θ_2. The function $\bar\theta(r_1,r)$ is the critical angle between the radii r_1 and r, and represents the angle traversed by a trip from r_1 whose path terminates at a tangent point to the circle of radius r (see mathematical note 2.1, p.24).

The forms of these functions derived from the radially symmetric model for Greater Manchester, 1965 are presented in figure 4.6, p.65.

Transport expenditure and urban economic theory

5.1 Introduction
The model of spatial interaction described in the previous chapter allows us to gain an important insight into the treatment of transport expenditures in economic theories of urban spatial structure. In this chapter we examine the works of several urban economists. These works contain several assumptions about the form of the transport-expenditure function, each based on the premise that all commuter-trip destinations are concentrated in the Central Business District[8]. We will show that these assumptions lead to contradictions in the theories, and that the assumption that all destinations are concentrated in the CBD needs to be relaxed in order to obtain a transport-expenditure function with the correct properties. A transport-expenditure function will be derived from the model as the average expenditure of car commuters on transport for each place of residence, and we shall show that this function possesses the required properties. Expenditure on transport in urban areas has been the major explanatory variables in spatial economic theories of the city. In particular, they have been considered to be the key to the determination of rents, densities, and land uses in the city. The basic observations underlying the role of transport expenditures in the determination of rents are that central locations in cities require low expenditure on transport and command high rents, while peripheral locations require more expenditure on transport and command lower rents.

In addition, densities are found to be high in areas of low transport expenditure and high rents, and low in areas of high transport expenditure and low rents.

These observations have existed in the economic literature for many years. Recent work in the field, involving the development of mathematical models of the urban land market, has been more concerned with attempting to integrate these observations into a consistent theory. Most prominent in this respect have been the works of Alonso (1965), Beckmann (1973), Mills (1967; 1969; 1972), Muth (1961; 1969), and Wingo (1961). All of these models attempt to explain the spatial structure of the city from economic assumptions coupled with some representation of urban transport expenditures. The economic assumptions made by these authors

---

[8] Some authors consider the effects on rent patterns of another competing centre of employment. A more general approach, which suggests a generalization to many centres of employment, is outlined by Alonso (1967, p.42). A more recent study by Evans (1973) furnishes evidence that such a generalization is necessary in order to explain the patterns of residential location in the Greater London Area. The detailed implications of this approach for the spatial distribution of transport expenses requires further exploration.

differ in several important aspects, and will not be reviewed here. In this chapter we wish to focus on their treatment of urban transport.

Clearly, all urban economists work in the Central Business District. The tradition followed by these economists is to assume that all employment is concentrated at a central location, which is usually represented as a point. All travel is then assumed to terminate at this single employment centre. This tradition owes its origin to the theory of agricultural production proposed by von Thünen.

In von Thünen's 'isolated' state all goods are sold at a single market centre, distance from that market being a major determinant of land rent. The assumption of a single employment centre has been maintained by the urban economists in the face of increasing decentralization and the decline of the role of the Central Business District as the single focus of productive activity. It has survived despite its decreasing relevance because it produces simplifications in the analysis and permits important results to be derived which would be difficult to obtain otherwise.

The relaxation of the assumption of a single employment centre poses several problems. When employment is distributed throughout the city, the expenditure on travel to work from a given residence is dependent on the choice of workplace. Our problem is to determine a transport-expenditure function given a distribution of workplaces and a representation of the transportation system. This transport-expenditure function must therefore be an average expenditure on travel to work, the average being weighted by the conditional distribution of workplaces corresponding to each place of residence.

In this chapter we derive this transport-expenditure function and show that it possesses the properties necessary to overcome the difficulties inherent in the view of transport postulated by the urban economists. It is important to note, however, that we have adopted time rather than money as the measure of transport expenditure. Since our analysis is restricted to car travel only, and does not consider public transport, travel time appears to be the best measure of the variable portion of transport expenditures. All other expenditures associated with transport: petrol, depreciation, maintenance, and the like, are closely related to it, and should therefore be represented with sufficient precision by travel time. Throughout the remainder of the analysis, our references to transport expenditures will not distinguish between money expenditures and time expenditures. In a similar manner, cost and time will be used interchangeably.

It is important to point out here that, while in the traditional economic analysis the transport-expenditure function is usually assumed to be given and the distribution of residences is to be derived, in this case the situation is reversed. The distribution of residences is assumed to be given, and the transport expenditure function is derived.

The spatial framework of the analysis is consistent with that of the urban economists, but differs from it in one important respect. We have preserved the view of the city as a continuous terrain with all the quantities under discussion regarded as functions of distance from a single centre. To this extent we follow their tradition. The assumption of radial symmetry and the display of all information as a function of distance from the city centre will be maintained throughout the analysis. However, we do not wish to retain the assumption that all employment is concentrated in the CBD. Thus, while the subject of the economists' discussion reduces to a single line, our discussion must encompass the two-dimensional plane of the city.

5.2. A critique of the urban economists' view of transport

Alonso, Mills, Muth, and Wingo are concerned with equilibrium in the city's land market. They all devote a major part of their work to the spatial structure of the housing market. In deriving the optimum location of households, they consider that households may be willing to pay more for rent when they have to pay less on transport, and less on rent when they have to pay more for transport. Muth and Mills derive the equilibrium location of households by equating, for a given quantity of housing, the marginal expenditures on rent and on transport. Mills deduces the form of the urban rent gradient by equating the value of the marginal productivity of land in transport with land rent.

Clearly, in all these models the form of the transport-expenditure function plays a crucial role. The different authors make different statements about it, although it can be shown that they all share very similar views. Given that all these authors consider workplaces to be concentrated at a single point in the city centre[9], we can infer that commuting costs at the centre are zero, since people living there do not have to spend anything on travel to work. For the same reason, commuting costs must increase with distance from the city centre. This requirement is stated clearly in all the works mentioned above.

Mills, Muth, and Wingo further assume that the rate of increase of the commuting costs decreases with distance from the city centre. This

[9] Mills attempts to avoid concentrating all workplaces in the city centre, and distributes some workplaces throughout the city. However, oddly enough, these workplaces where output is produced do not require commuter transportation. Mills states that "the total demand for transportation at u (distance from the city centre) is proportional to the output produced beyond u. When output is goods, the assumption can be interpreted to mean that a certain fraction of goods is transported to the city centre. When output is housing services, the assumption can be interpreted to mean that a certain amount of commuter traffic is generated per household" (Mills, 1969, p.243). Mills appears to ignore the demand for transportation by commuters who produce output outside the city centre, although output is produced throughout the city. From a transportation point of view, then, this assumption is identical with that of Alonso and the other economists.

assumption is based on the previous assumption and the observation that congestion in central areas reduces speeds and increases costs per unit distance. In Mills' words (Mills, 1967, p.204):

"travel is inevitably slower in denser, higher rent areas, even in an optimum transportation system".

A similar statement is made by Muth (1969, pp.19–20):

"If anything, traffic generally moves more rapidly at greater distances from the CBD; less time in transit is thus spent per mile and vehicles, up to a point, operate at more economical speeds with less stopping and starting. Therefore, I assume, that the variable portion of transport costs increases at a non-increasing rate with distance from the CBD or any other centre of activity."

Wingo is more explicit and provides a stricter form for the transport-expenditure function (Wingo, 1961, pp.96–97):

"We can summarise the technical conditions by describing the common transportation function as in Figure 26 (reproduced here as figure 5.1). In this function FH we have assumed a constant operating velocity represented by the constant slope FH, and an average ingression loss, all of which occurs at the centre, represented by OF. An alternate assumption would distribute ingression losses throughout the system, yielding a function such as OGH, which implies a decline in average velocity as one moves from G into increasing congestion until the centre is reached. It is the OGH type of function which will be assumed hereafter".

Wingo later remarks (Wingo, 1961, p.101):

"The curve OGH is probably more representative of the transportation function as generally experienced in urban areas".

It can easily be seen that the general form of the transport-expenditure function described by Wingo is implied in the analyses of the other economists. Since all workplaces have been assumed to be concentrated at the city centre, the function must have a value of zero at the centre. The people living and working at the centre should not be spending anything on transport. The transport-expenditure function then increases

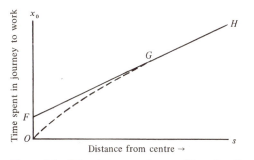

Figure 5.1. Wingo's transport-expenditure function (from Wingo, 1961, p.97).

at a nonincreasing rate with distance from centre, since its gradient is the cost per unit distance, which is assumed to decrease with distance from the centre. It must also be asymptotic to a straight line with a positive slope. This follows since velocities cannot increase indefinitely with current transport technology and are therefore bounded from above. After a certain distance from the city centre, velocities must reach a constant value. Cost per unit distance must therefore approach a constant gradient. The gradient must be strictly positive because the cost per unit distance cannot be zero. Since all employment is assumed to be at the centre there must always be an additional cost of travel when distance from the centre increases.

Wingo's form of the transport-expenditure function can be taken as the form implicit in the work of the other economists, being derived directly from their assumptions. This form of the function is consistent, in the case where everyone is assumed to be travelling on radial routes to the city centre, with our observations using Manchester data for car commuters in 1965.

The expenditure on travel from any point to the city centre was obtained by integrating the inverse velocity field for Greater Manchester (figure 2.8) along a radial path. This integration yields expenditure on travel to the city centre as a function of distance from the centre. The resulting function is illustrated by the curve shown in figure 5.2.

As can be seen by comparing figure 5.1 and 5.2, both curves share the same properties. Both are zero at the origin; both increase at a decreasing rate; and both are asymptotic to a straight line with a positive gradient.

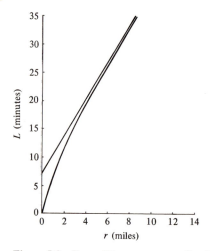

Figure 5.2. Expenditure on commuting to the centre of Manchester as a function of distance of a given residence from the centre, 1965.

We now proceed to analyse the difficulties inherent in this form of the transport-expenditure function, and hence also in the assumption that all destinations are concentrated in the Central Business District. Wingo (1961, p.65) makes the assumption that the sum of expenditures on rent and transport is fixed for any given household. Muth (1969, p.21) deals with the total household budget, rather than with a special budget for housing and transport. Muth postulates a composite good which includes all goods purchased except housing and transport. The household in this case is assumed to make its location decision by maximizing a utility function which depends on the quantities of housing and the composite good purchased, subject to a household-budget constraint.

In mathematical notation, we maximize $U(q, z)$ subject to the budget constraint

$$y = p_z z + p(r)q + L(r) \,,$$

where $p(r)$ is the price of a unit of housing (or land) at a distance r from the city centre, q is the quantity of housing (or land) purchased, $L(r)$ is the expenditure on travel to the centre for a household locating at r, z is the quantity of the composite good, and p_z is the price of a unit of the composite good. A necessary condition for household equilibrium then is

$$\frac{\mathrm{d}L}{\mathrm{d}r} = -q_0 \frac{\mathrm{d}p}{\mathrm{d}r} \,,$$

where $\mathrm{d}p/\mathrm{d}r$ and $\mathrm{d}L/\mathrm{d}r$ are the rates of change of house prices and transport expenditures with respect to distance from the city centre, and q_0 is the equilibrium quantity of housing. This condition, which appears in Muth (1969, p.22) and Mills (1967, p.205), requires that the household cannot increase its real income by a change of location. It states that, for a given quantity of housing, the rate of change of housing expenditure is of the same magnitude as the rate of change of transport expenditure, but with the opposite sign. Muth then argues that, since transport expenditure increases with distance from the centre, house prices must decline with distance. According to figures 5.1 and 5.2 the transport-expenditure gradient approaches a strictly positive value at large distances from the centre. If we now assume that the quantity of housing purchased is bounded above, namely that people cannot buy indefinitely large houses, it follows that house prices must decline at a rate bounded away from zero. In other words, it would appear that the house price or the rent function should approach a straight line of strictly negative slope. From these conditions we can deduce that house prices or rents should become negative at some distance from the centre. This conclusion is deduced directly from the first-order condition for equilibrium stated above, and applies, therefore, to the analysis of Mills as well[10].

In the case of Wingo, where the sum of expenditures on rent and transport expenditures is assumed to be fixed, the conclusion is similar. Since transport expenditures increase indefinitely, rents are bound to become negative at some point.

Alonso's analysis (Alonso, 1965) is similar to that of Muth but is more general, since Alonso assumes that distance from the city centre has a disutility associated with it. The utility function in this case takes the form

$$U(q, z, r) \ .$$

The equilibrium condition mentioned above does not apply in this case. However, the conclusion that land rents become negative at large distances from the city centre still applies, unless the marginal utility of distance from the city centre is positive, namely unless people derive increasing utility from moving further out, which according to Alonso is unlikely (Alonso, 1965, pp.34–35).

The occurrence of negative rents is usually obscured by introducing a 'city limit' into the economic analysis. Urban economists then restrict their attention to the region within this limit, leaving the rest of the country to students of another field. However, this practice raises the problem of specifying the appropriate boundary conditions. In particular we should not expect the house-price gradient to change abruptly at some arbitrary distance from the centre, nor is there data to support such a contention. It is more reasonable to expect house prices to approach a constant positive value. This, however, would not be consistent with the economists' transport-expenditure function, because it would require the latter to approach a positive constant. If the transport-expenditure function is to approach a constant at large distances from the city centre, we must abandon the requirement that all destinations are concentrated in the CBD. At large distances from the CBD the price of housing or land must be unrelated to the cost of travelling to the city centre.

---

[10] Muth later modifies his analysis (Muth, 1969, p.72) and assumes that transport expenditures increase at a *constant* rate, in order to derive a negative exponential house-price gradient. This would imply that transport expenditures must vary linearly with distance from the city centre. Since all jobs are located in the CBD, according to his analysis, travel cost must be zero at the centre. If we assume that the gradient of the transport-expenditure function is positive, rents become negative at large distances from the centre. It must follow that the gradient of the transport-expenditure function is zero, and hence that transport expenditures would be zero everywhere. In the light of these criticisms, Muth's derivation of the negative exponential house-price function requires some revision.

5.3. The transport-expenditure function

It appears that the economic analysis of cities which requires that all workplaces be located in the Central Business District leads to unrealistic conclusions about the structures of rents and house prices. We must therefore reject the assumption that all workplaces are located in the Central Business District and consider a distribution of jobs over the entire city. With such a distribution it becomes possible to obtain a more realistic transport-expenditure function. In particular a person residing at a distant location could obtain employment close to his place of residence, and is therefore not required to bear the large transport expenditure involved in commuting to the CBD. When this is the case, transport expenditures no longer increase indefinitely with distance from the centre but reach a maximum value.

With a distribution of jobs, transport expenditures are not uniquely determined at each distance from the city centre. Since people work everywhere, transport expenditures at a given location can only be determined as an average expenditure for people living there, and working in various destinations throughout the city. When these expenditures are averaged over many workplaces, they are no longer zero in the city centre. Expenditures on travel to work from the centre are obtained by averaging expenditures for people living in the centre and working throughout the city, and can no longer be taken as negligible.

Furthermore, when the analysis is restricted to car travel, excluding public transport, central locations are not necessarily locations of least transport expenditures, as in the preceding economic analysis. Given that people living in the centre must travel through congested areas on their way to work, it is no longer necessarily true that central locations are most accessible to jobs. All of these properties are exhibited in the transport-expenditure function derived for Greater Manchester, 1965. This function was obtained, as stated earlier, by taking as given the distribution of residences and workplaces of car commuters in Greater Manchester, and by allocating commuters between residences and workplaces in accordance with the continuous trip-distribution model. The transport-expenditure function was derived from the model as an average expenditure on transport for commuters residing in a given location. The total expenditure on transport at that location was obtained by summing the cost of all trips originating in that location. The average expenditure on transport was obtained by dividing that total by the number of origins at that location. (The analytical expression for the transport-expenditure function appears in mathematical note 5.1, p.84.)

Figure 5.3 illustrates the transport-expenditure function for Greater Manchester, 1965. The vertical axis represents commuting costs, measured in this case in minutes. The horizontal axis represents distance from the city centre.

As can be seen from figure 5.3, as one proceeds outwards from the city centre, transport expenditures *decrease* for the first half mile from 9·5 minutes to about 8·8 minutes. This effect is associated with congestion in the central area, since it did not appear when average transport expenditures were derived for a uniform travel-cost field. Transport expenditures then increase, *at an increasing rate*, until they reach a point of inflection about 4·5 miles out. They subsequently increase *at a decreasing rate* to reach a value of about 20 minutes at a distance of 14 miles from the centre. Transport expenditures then level out and remain at this value for the rest of the range. The latter effect is probably due to the predominance of local jobs at large distances from the centre. It is also important to note that transport expenditures in general are far lower, when only radial travel to the city centre is considered, than those presented earlier (figure 5.2). This can be easily seen by comparing figures 5.2 and 5.3, which are drawn to the same scale. The latter is markedly shallower.

It must follow, therefore, that the assumption that all commuters travel to the CBD, overestimates their expenditures on transport. This can be best illustrated by comparing transport expenditures for people who actually work in the Central Business District with those who work elsewhere. Such a comparison is shown in figure 5.4.

The form of figure 5.4 was obtained in a similar manner to the preceding figure 5.3. The total expenditure on travel to work in a given location was obtained by summing the costs of all trips terminating in that location. The average expenditure on transport for people working in that location was obtained by dividing that total by the number of destinations at that location. (The analytical expression for the work-based transport-expenditure function appears in mathematical note 5.1, p.85.) Figure 5.4 exhibits the average expenditures on transport for people *working* at a given distance from the city centre.

As can be seen from this figure, within most of the urban area, those car commuters who work in the city centre spend most on travel to work.

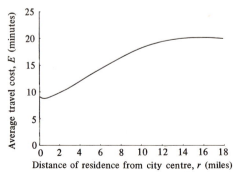

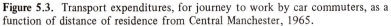

Figure 5.3. Transport expenditures, for journey to work by car commuters, as a function of distance of residence from Central Manchester, 1965.

Expenditure on travel to work declines steadily for more suburban workplaces and reaches a minimum at a distance of approximately 7 miles from the centre. Employees working at these distances spend the least, on average, on commuting. Jobs located further out require higher expenditures on transport, and the function did not seem to level out within the range of study[11].

Clearly, then, the economists' assumption that jobs are concentrated at the centre considerably overestimates the amount of time or money spent on commuting. Most people will have smaller expenditures on commuting than those postulated by the economists. We must, therefore, conclude that the economists will inevitably attribute more to the effects of transport expenditures on rent and location than is, in fact, attributable to them when a distribution of jobs throughout the urban region is taken into account[12].

We return to the transport-expenditure function exhibited in figure 5.3. The function possesses the correct properties at large distances from the centre: it reaches a maximum and does not go on increasing indefinitely. Near the centre, within the first half mile range, it does not agree with the economists' assumptions because it is decreasing rather than increasing

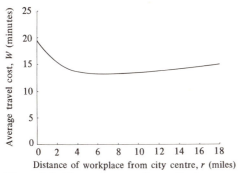

Figure 5.4. Transport expenditure, for journey to work by car commuters, as a function of distance of workplace from Central Manchester, 1965.

[11] Note the resemblance of figure 5.4 to the accessibility to residences curve in figure 4.5, and the resemblance of figure 5.3 to the accessibility to jobs curve in the same figure. Both pairs of curves seem to share the property that their derivatives are of opposite signs throughout the range. The minimum transport-expenditure location corresponds to the maximum accessibility location. This resemblance suggests a relation between transport expenditures and accessibility, which requires further study.

[12] The high cost of commuting to the centre gives us an indication of the forces of decentralization at work in modern cities. From the labour force's point of view, savings accrue when jobs can be found away from the city centre. We must therefore expect the number of vacancies in central areas to be higher, and wages offered to be higher than those in outlying locations. At large distances from the centre, workplaces again require large expenditures on commuting. In this case employees must come from distant locations because of the low density of the population.

there. If we assume that rents are at a maximum in the city centre, due
to competition from commercial and other uses which require more
central locations, then the necessary condition for household equilibrium
is not satisfied in this range. Both rents and transport expenditures have
negative gradients. We must conclude, then, that no household would
locate within this range. Remembering that we have restricted our analysis
to car users, we can rephrase this prediction by saying that no resident
living within this range would commute to work by car. This prediction is
clearly borne out by our observations in Manchester. There are no car
commuters within this distance from the centre. This is illustrated by the
gamma distribution fitted for the residences of car commuters (figure 4.3,
p.60) which has the property that the density of residences at the centre
is zero. At large distances from the centre, if this equilibrium condition is
to be satisfied, we should expect the rent gradient to level out when the
transport gradient levels out. We have no *a priori* reason, however, to
expect it to level off at zero, as in the negative exponential function
derived in Muth (1969, p.72) and Mills (1969, p.245).

The derived transport-expenditure function thus differs from the function
postulated by the urban economists in several important ways. It is positive,
rather than zero at the origin. It decreases for a short range near the origin,
rather than increasing everywhere. It generally increases at a slower rate.
Its slope first increases then decreases over the range, rather than being
nonincreasing everywhere. And finally, the function approaches a constant
value rather than increasing indefinitely. All of these properties can be
justified on quite simple grounds and are consistent with the observations.

It must be noted, however, that these properties are exhibited by
functions derived for one particular urban area. We have, therefore,
confined ourselves to the discussion of their general features, and not
discussed the specific values obtained. More cases need to be examined
for these results to be more firmly established. It appears that future
analysis would have to consider a more appropriate form for the transport-
expenditure function than that previously postulated in urban economic
theories.

Mathematical note 5.1. *The derivation of the transport-expenditure*
function

The transport-expenditure function is derived from the continuous model
of spatial interaction. The transport expenditure, $E(r_1)$, at a distance r_1
from the city centre is the average expenditure of the residents in a small
area there on transport. It is obtained by summing the costs of individual
trips and dividing this sum by the total number of trips from a given place
of origin

$$E(r_1) = \frac{1}{O(r_1)} \int_0^\infty \int_0^{2\pi} \int_0^{2\pi} T(r_1,\theta_1,r_2,\theta_2) t(r_1,\theta_1,r_2,\theta_2) r_2 \, d\theta_1 \, d\theta_2 \, dr_2 \; ,$$

$$(5.1)$$

where $O(r_1)$ is the density of origins at r_1, $T(r_1,\theta_1,r_2,\theta_2)$ is the density of trips between a small area at (r_1,θ_1) and a small area at (r_2,θ_2), and $t(r_1,\theta_1,r_2,\theta_2)$ the travel time between these points. The form of the transport-expenditure function $E(r)$ derived for Greater Manchester, 1965, is presented in figure 5.3.

In a similar manner we obtain the work-based transport expenditure, $W(r_2)$, as the average expenditure on transport for people working in a given location at a distance r_2 from the city centre.

$$W(r_2) = \frac{1}{D(r_2)} \int_0^\infty \int_0^{2\pi} \int_0^{2\pi} T(r_1,\theta_1,r_2,\theta_2) t(r_1,\theta_1,r_2,\theta_2) r_1 \, \mathrm{d}\theta_1 \, \mathrm{d}\theta_2 \, \mathrm{d}r_1 ,$$

(5.2)

where $D(r_2)$ is the density of destinations at r_2. The form of this function, derived for Greater Manchester, 1965, is presented in figure 5.4.

6

Concluding remarks

The geometry of movement developed in this study and applied to various geometrical theories of urban and regional space makes it possible to treat movement and transport in a more realistic framework without surrendering the basic geometrical approach.

Several assumptions which were made throughout the analysis, particularly the assumption that the transport background is given as a smooth function defined over the Euclidean plane, and the assumption of the radial symmetry of urban regions, need to be further explored. The point, however, is not to relax these assumptions on the grounds that they are not real, but to continue to investigate the advantages and disadvantages of retaining them in the analysis of specific problems. It may well be that additional complication will not result in additional insights.

It seems that many interesting problems can be explored further without relaxing these assumptions. In particular, several extensions of the continuous model of spatial interaction merit further study.

In order to deal successfully with the problem of capacity restraint we have to incorporate into the model the spatial distribution of roads. We can then formulate the relationships between flows, velocities, and road capacity in a continuous framework. With the incorporation of capacity-restraint algorithms it is possible to study several questions concerning the effects of changes in distributions on the performance of the transport system, on the distribution of residences and workplaces, on the distributions of accessibilities and transport expenditure, and possibly, on the distribution of incomes and benefits or costs over geographical space.

Several other models, similar to the trip-distribution model developed here, which were previously restricted to a network representation of space could be further explored in a geometrical context. Their study in this context may reveal many of the relationships between spatial variables which still remain hidden.

Generally, it can be said, that variables which can be conceived as distributions over geographical space, and the intricate relationships between these variables and human behaviour affecting them and affected by them offer many intriguing questions which still remain unanswered.

It is hoped that this study has shed some light on existing geometrical theories of urban and regional space, and has opened the way for other possible theoretical advances by integrating concepts and techniques for regional analysis into the geometrical framework.

References

Alonso, W., 1965, *Location and Land Use* (Harvard University Press, Cambridge, Mass.).

Alonso, W., 1967, "A reformulation of classical location theory and its relation to rent theory", *Papers and Proceedings of the Regional Science Association,* **19**, 23-44.

Angel, S., Hyman, G., 1970, "Urban velocity fields", *Environment and Planning,* **2**, 211-224.

Angel, S., Hyman, G., 1971, "Urban travel time", *Papers and Proceedings of the Regional Science Association,* **26**, 85-99.

Angel, S., Hyman, G., 1972a, "Urban spatial interaction", *Environment and Planning,* **4**, 99-118.

Angel, S., Hyman, G., 1972b, "Urban transport expenditures", *Papers and Proceedings of the Regional Science Association,* 105-123.

Angel, S., Hyman, G., 1972c, "Transformation and geographic theory", *Geographical Analysis,* **4** (4), 350-367.

Beckmann, M., 1973, "Equilibrium models of residential land use", *Regional and Urban Economics,* **3** (4), 361-368.

Berry, B., Simmons, J., Tennant, R., 1963, "Urban population densities: structure and change", *Geographical Review,* **29**, 389-405.

Broadbent, T. A., 1970, "Notes on the design of operational models", *Environment and Planning,* **2**, 469-476.

Bunge, W., 1964, "Patterns of location", Discussion Paper 3, Michigan Inter-University Community of Mathematical Geographers, Ann Arbor.

Bunge, W., 1966, *Theoretical Geography,* Lund Studies in Geography, series C, number 2 (G. W. Gleerup, Lund, Sweden).

Bussière, R., Snickars, F., 1970, "Derivation of the negative exponential model by an entropy maximising method", *Environment and Planning,* **2**, 295-301.

Casetti, E., 1967, "Urban population density patterns: an alternate exploration", *Canadian Geography,* **11** (2), 96-100.

Christaller, W., 1966, *Central Places in Southern Germany,* translated by C. W. Baskin (Prentice-Hall, Englewood Cliffs, NJ).

Clark, C., 1951, "Urban population densities", *Journal of the Royal Statistical Society,* **114**, 490-496.

Courant, R., Hilbert, D., 1953, *Methods of Mathematical Physics,* volume 1 (Interscience, New York).

Dacey, M. F., 1968, "A model for the areal distribution of population in a city with multiple population centres", *Tijdschrift voor Economische en Sociale Geographie,* **59**, 232-236.

Evans, 1973, *The Economics of Residential Location* (Macmillan, London).

Feller, W., 1968, *An Introduction to Probability Theory and its Applications,* volume 1 (John Wiley, New York).

Glasgow Corporation, 1965, *A Highway Plan for Glasgow,* prepared by Scott, Wilson, Kirkpatrick and Partners, Glasgow.

Gurevich, B. L., Saushkin, Yu. G., 1966, "The mathematical method in geography", *Soviet Geography,* **7** (2), 3-35.

Haggett, P., 1965, *Locational Analysis in Human Geography* (Edward Arnold, London).

Haggett, P., Chorley, R. J., 1969, *Network Analysis in Geography* (Edward Arnold, London).

Hall, P., Drewett, R., Gracey, H., Thomas, R., 1973, *The Containment of Urban England* (Allen and Unwin, London).

Harvey, D., 1969, *Explanation in Geography* (Edward Arnold, London).

Hilbert, D., Cohn Vossen, S., 1952, *Geometry and the Imagination* (Chelsea, New York).

Holroyd, E. M., 1966, "Theoretical average journey lengths in circular towns with various routing systems", Road Research Laboratory, report number 43, Crowthorne, Berks.

Hoyt, H., 1940, "Urban decentralisation", *Journal of Land and Public Utility Economics*, **16**, 270-276.

Huygens, C., 1912, *Treatise on Light,* translated by S. P. Thompson (Macmillan, London).

Isard, W., 1956, *Location and Space Economy*, Regional Science Studies Series (MIT Press, Cambridge, Mass.).

Isard, W., Liossatos, P., 1972a, "On location analysis for urban and regional growth situations", *Annals of Regional Science*, **6**, 1-27.

Isard, W., Liossatos, P., 1972b, "Social energy: a relevant new concept for regional science?", in *London Papers in Regional Science 3*, Ed. A. G. Wilson (Pion, London), pp.1-30.

Jenks, G. F., 1963, "Generalization in statistical mapping", *Annals of the Association of American Geographers*, **53**, 15-26.

Kirby, H. R., 1970, "Normalising factors for the gravity model—an interpretation", *Transportation Research*, **4**(1), 37-50.

Lam, T. N., Newell, G. F., 1967, "Flow dependent traffic assignment in a circular city", *Transportation Science*, **1**(4), 318-361.

Lane, R., Powell, T. J., Smith, P. P., 1971, *Analytical Transport Planning* (Duckworth, London).

Lösch, A., 1967, *The Economics of Location*, translated by W. H. Woglom (John Wiley, New York).

Lowry, I. S., 1968, "Seven models of urban development: a structural comparison", Urban Development Models, Special Report 97, Highway Research Board, Washington, DC, pp.121-163.

Miehle, W., 1958, "Link-length minimization in networks", *Operations Research*, **6**(2), 232-243.

Mills, E. S., 1967, "An aggregative model of resource allocation in a metropolitan area", *American Economic Review*, **57**(2), 197-210.

Mills, E. S., 1969, "The value of urban land", in *The Quality of the Urban Environment*, Ed. H. Perloff (Resources for the Future, Washington, DC).

Mills, E. S., 1972, *Studies in the Structure of the Urban Economy* (Resources for the Future, Baltimore).

Muth, R., 1961, "The spatial structure of the housing market", *Papers and Proceedings of the Regional Science Association*, **7**, 207-220.

Muth, R., 1969, *Cities and Housing* (University of Chicago Press, Chicago).

Newling, B. E., 1969, "The spatial variation of urban population densities", *Geographical Review*, **59**(2), 242-252.

Niedercorn, J. H., 1971, "A negative exponential model of urban land use densities and its implications for metropolitan development", *Journal of Regional Science*, **11**(3), 317-326.

Owens, D., 1968, "Estimates of the proportion of space occupied by roads and footpaths in towns", report number LR 154, Ministry of Transport and Road Research Laboratory, Crowthorne, Berks.

Park, E. R., Burgess, E. W., 1925, *The City* (University of Chicago Press, Chicago).

Rhind, D. W., 1971, "Automatic contouring—an empirical evaluation of some differing techniques", *Cartographic Journal*, **8**(2), 145-158.

SELNEC, 1968, "Study Area Characteristics", Technical Working Paper 4, South East Lancashire and North East Cheshire Transportation Study, Manchester.

Shannon, C. E., Weaver, W., 1964, *The Mathematical Theory of Communication* (University of Illinois Press, Urbana).

Spurkland, S., 1967, "Mathematical tools for urban studies", reprint number 14, Norwegian Institute of Urban and Regional Research.

Stewart, J. Q., Warntz, W., 1968, "Physics of population distribution", in *Spatial Analysis: A Reader in Statistical Geography*, Eds. B. J. L. Berry, D. F. Marble (Prentice-Hall, Englewood Cliffs, NJ), pp.130-146.

Tobler, W., 1963, "Geographic area and map projection", *Geographical Review*, **53**, 59-78 (reproduced here, page 135).

Tobler, W., 1966, "Medieval distortions: the projections of ancient maps", *Annals of the Association of American Geographers*, **56** (2), 351-360.

Tomlin, J. A., Tomlin, S. G., 1968, "Traffic distribution and entropy", *Nature*, **220**, 974-976.

von Thünen, J. H., 1966, *Von Thünen's Isolated State*, Ed. P. Hall (Pergamon Press, Oxford).

Wagon, D. J., Hawkins, A. F., 1970, "The calibration of the distribution model for the SELNEC study", *Transportation Research*, **4** (1), 103-112.

Wardrop, J. G., 1952, "Some theoretical aspects of road traffic research", *Proceedings of the Institute of Civil Engineers*, **1** (1), 325-378.

Wardrop, J. G., 1969, "Minimum cost paths in urban areas", *Beiträge zur Theorie des Verkehrsflusses, Strassenbau und Strassenverkehrstechnik*, **86**, 184-190 (reproduced here, page 155).

Warntz, W., 1965, "A note on surfaces and paths and applications to geographic problems", Discussion Paper 6, Michigan Inter-University Community of Mathematical Geographers, Ann Arbor, Michigan (reproduced here, page 121).

Warntz, W., 1967, "Global science and the tyranny of space", *Papers and Proceedings of the Regional Science Association*, **19**, 7-19 (reproduced here, page 108).

Williams, H. C. W. L., Ortuzar-Salas, J. D., 1974, "Some generalisations and applications of the velocity field concept 1. Trip patterns in idealised cities", WP 30, Institute of Transport Studies, University of Leeds, England.

Williams, H. C. W. L., 1974a, "Some generalisations and applications of the velocity field concept 2. Problems relating to traffic restraint, optimal investment and velocity field dynamics", WP 31, Institute of Transport Studies, University of Leeds, England.

Williams, H. C. W. L., 1974b, "Some generalisations and applications of the velocity field concept 3. The location and pricing of park-and-ride in an urban area", WP 32, Institute of Transport Studies, University of Leeds, England.

Wilson, A. G., 1967, "A statistical theory of spatial distribution models", *Transportation Research*, **1**, 253-269.

Wilson, A. G., 1970, *Entropy in Urban and Regional Modelling* (Pion, London).

Wilson, A. G., Hawkins, A. F., Hill, G. J., Wagon, D. J., 1969, "Calibrating and testing the SELNEC transport model", *Regional Studies*, **3**, 337-350.

Wingo, L., 1961, *Transportation and Urban Land* (Resources for the Future, Washington, DC).

Appendix

A computer program for the model

This program was written in order to calibrate the continuous and radially symmetric model of spatial interaction described previously. The program uses as inputs the velocity field for an urban area, the density functions for the origins and destinations of commuter trips, and the average travel time in the region. No other inputs are required.

By using the procedure described in chapter 2, the travel time between pairs of points in the velocity field is computed. These travel times are then used, together with the other inputs, to solve the equations of the model by iterative methods. The solutions to these equations produce forms for the balancing factors, and a value for the parameter μ. We can thus determine the trip density function, and hence the density of trips between pairs of locations.

The results of the model calibration are then used to produce the predicted travel-time distribution, the normalizing factors which are interpreted as accessibilities, the cordon-crossing functions which describe the spatial distribution of traffic flows, and the average expenditure on transport at each location.

To facilitate understanding, the program has been divided into subroutines. The relationships between these subroutines are illustrated in figure A1.

Subroutine TIMES computes the array of travel times corresponding to each (r_1, r_2) pair, for each value of the angular difference A. The radial components are stored in a triangular array such that only values for which $r_1 \leqslant r_2$ need be computed or stored, the remaining values being determined by symmetry. If either r_1 or r_2 is zero, or if the angular difference is zero, then we are dealing with a radial trip. In this case the travel time can be computed by a single integration, which is performed in TRADIAL. Otherwise, control is passed down to TGEN. TGEN is basically the procedure described in urban travel time. The first decision to be made is whether we are dealing with a direct trip or a through trip. This is determined in DOTKBAR by comparing the values of the angular difference and the critical angle corresponding to the (r_1, r_2) pair under consideration. This information is passed back to TGEN by means of the parameter DOT. The maximal characteristic KBAR corresponding to (r_1, r_2) is also passed back. TGEN then calls the subroutine KDIRECT for a direct trip, and the subroutine MINRAD for a through trip.

KDIRECT computes the characteristic of the direct route between two points with specified radii and angular differences. An iterative procedure is used, starting with an initial guess based on a constant velocity field. Given each estimate of the characteristic, the angular difference this would imply is computed by the subroutine ANGLE. The characteristic is then recalculated until it converges on the correct angular difference. However, for pairs of points for which the angular difference is close to the critical angle, difficulties can arise. An approximation for the characteristic could exceed the maximal characteristic corresponding to the (r_1, r_2) pair. If this happens ANGLE would be unable to produce a real angular difference and the program would come to a grinding halt, signalling an error message. To help overcome the problem the maximal characteristic KBAR has been passed down from TGEN. Each approximation to the characteristic is compared to KBAR and, if it is found to exceed KBAR, it is set to a smaller value. This value is indicated by the variable SPACE. The value set for SPACE was decided by experiment, and represents a compromise between accuracy of results and speed of computation, as of course do the convergence limits. After the characteristic has been determined control is passed, via TGEN, to the subroutine TDIRECT, which computes the travel time integral, as described in urban travel time.

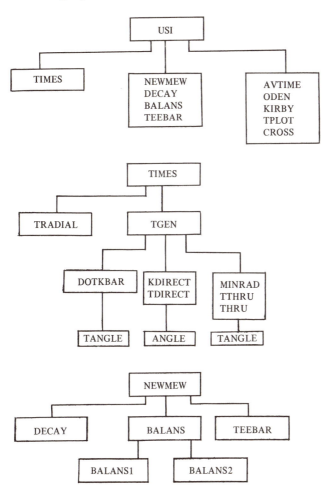

Figure A1. The structure of the program.

MINRAD computes the minimum radius RMIN of the through route between two points with specified radii and angular differences. The procedure is essentially the same as in KDIRECT but the method must be modified since a through route passes through its minimum radius. Hence, after the initial guess for RMIN, the corresponding angle must be computed by dividing the range of integration into two sections. The integration is computed in TANGLE, which computes the critical angle between the approximation to the minimum radius and the radius of the point under consideration. The minimum radius is then recalculated and the process repeated until we converge on the correct angular difference. A related difficulty to that occurring in KDIRECT arises here, but is is much simpler to deal with. Since the characteristic can be directly calculated from the minimum radius it is only necessary to check that each estimate for RMIN is less than both radii under consideration. After RMIN has been determined, control is passed, via TGEN, to the subroutine TTHRU which divides the range of integration into the two appropriate sections and calls THRU, which computes the travel time along each section. When the angular difference approaches 180° the minimum radius is small and the above procedure becomes very inefficient. Consequently the travel time for these trips is determined by linear extrapolation.

The function of subroutine NEWMEW is to compute the value of the decay parameter μ which corresponds to the observed average travel time $\bar{t}$. The method adopted is equivalent to the method described in chapter 4, modified to deal with continuous spatial variables. An initial estimate for μ was taken as $\mu_0 = 2/\bar{t}$. It turned out that this was a very good approximation, the true value being $1 \cdot 96/\bar{t}$. It is interesting to compare this result with the value found in the SELNEC study for the product of the decay parameter and the average cost of travel—using a discrete formulation they obtained a value of about $2 \cdot 5$ (for calibration of the distribution model for the SELNEC study see Wagon and Hawkins, 1970).

After the initial estimate μ_0 is obtained, the subroutines DECAY and BALANS are called to determine the balancing factors $A(r_1)$, $B(r_2)$ implied by the value μ_0. These estimates for the balancing factors enable an initial estimate for the trip density function to be calculated, and hence the implied average travel time $\bar{t}_0$ can be computed, this being performed by the subroutine TEEBAR. Control then reverts to NEWMEW where a revised estimate for μ is made according to the formula $\mu_1 = \mu_0 \bar{t}_0 / \bar{t}$ and the above process repeated to produce a revised approximation $\bar{t}_1$ to the average travel time. After two estimates for μ and $\bar{t}$ have been made subsequent estimates are computed by linear interpolation, according to the formula

$$\mu_{n+1} = \frac{(\bar{t} - \bar{t}_{n-1})\mu_n - (\bar{t} - \bar{t}_n)\mu_{n-1}}{\bar{t}_n - \bar{t}_{n-1}}, \qquad n \geqslant 1 .$$

The above process is repeated until we obtain a value of $\bar{t}_n$ sufficiently close to the observed value $\bar{t}$.

The separation of the computation of the balancing factors into the subroutines DELAY and BALANS was adopted to avoid having to duplicate the large storage capacity needed for the travel-time matrix to construct the decay function. The role of DECAY is to perform the integration of the decay function with respect to the angular difference variable and store the result in an array **F** of reduced size. A similar integration is performed where the decay function is weighted by the travel time, to produce on array **G**. BALANS then computes the balancing factors from the array **F** by calling BALANS 1 and BALANS 2 alternately. The computation is stopped when the largest absolute difference between successive values of the destination balancing factor is sufficiently small. TEEBAR then uses the computed balancing factors and the array **G** to produce the average travel time.

The subroutine ODEN is a check on the accuracy of the balancing factors. The densities of residences and workplaces implied by the final values for the balancing factors are computed and compared to their original values. The subroutines AVTIME, KIRBY, and CROSS calculate the functions described in chapters 4 and 5. TPLOT computes the travel-time distribution.

Program USI

```
      PROGRAM USI(INPUT,OUTPUT,TAPE5=INPUT,TAPE6=OUTPUT)
      REAL MEW
      DIMENSION A(41),B(41),F(41,41),G(41,41),O(41),C(21),X(21),
     1TSTAR(60),ASTAR(41),BSTAR(41),AVTIME1(41),AVTIME2(41),TIME(861,19)
     2,T(60),A1(41),B0(41),B1(41)
      READ (5,1) TBAR,N1,N2,N3,N4,H1,H3,EPS1,EPS2,EPS3,EPS4,TRIPS
1     FORMAT(F6.3,4I2,F3.1,F9.7,4F5.3,F9.2)
      WRITE(6,9) TBAR,N1,N2,N3,N4,H1,H3,EPS1,EPS2,EPS3,EPS4,TRIPS
9     FORMAT(F6.3,4I2,F3.1,F9.7,4F5.3,F9.2)
      CALL TIMES(N1,N2,N3,H1,EPS1,EPS2,TIME)
      CALL SECOND(CP)
      WRITE (6,8) CP
      CALL NEWMEW(TBAR,N1,N2,H1,EPS3,EPS4,TRIPS,TIME,MEW,N3,A,B,F,G,A1,B
     10,B1)
      CALL SECOND(CP)
      WRITE (6,8) CP
      CALL DECAY(N1,N2,N3,MEW,TIME,F,G)
      CALL BALANS(N1,N2,H1,EPS3,F,A,B,A1,B0,B1)
      CALL TEEBAR(N1,N2,H1,TRIPS,A,B,G,TBAR0)
      CALL SECOND(CP)
      WRITE (6,8) CP
      CALL AVTIME(N1,N2,H1,A,B,G,AVTIME1,AVTIME2,MEW,TRIPS)
      CALL SECOND(CP)
      WRITE (6,8) CP
      CALL ODEN(N1,N2,H1,F,A,B,O)
      CALL SECOND(CP)
      WRITE (6,8) CP
      CALL KIRBY(N1,N2,H1,TRIPS,A,B,ASTAR,BSTAR,F)
      CALL SECOND(CP)
      WRITE (6,8) CP
      CALL TPLOT(N1,N2,N3,H1,MEW,A,B,TIME,TSTAR,T)
      CALL SECOND(CP)
      WRITE (6,8) CP
      CALL CROSS(N1,N2,N3,N4,H1,H3,MEW,TIME,F,A,B,C,X)
8     FORMAT(1H0,3HCP=,F10.5)
      STOP
      END
```

Subroutine NEWMEW

```
      SUBROUTINE NEWMEW(TBAR,N1,N2,H1,EPS3,EPS4,TRIPS,TIME,MEW,N3,A,B,F,
     1G,A1,B0,B1)
      DIMENSION A(N1),B(N2),F(N1,N2),G(N1,N2),TIME(861,19),A1(N1),B0(N2)
     1,B1(N2)
      REAL MEW,MEW0,MEW1,MEW2,K,K0,K1,K2
      MEW0=2.0/TBAR
      WRITE (6,5) MEW0
5     FORMAT(1H1,5HMEW0=,F8.5)
      CALL DECAY(N1,N2,N3,MEW0,TIME,F,G)
      CALL BALANS(N1,N2,H1,EPS3,F,A,B,A1,B0,B1)
      CALL TEEBAR(N1,N2,H1,TRIPS,A,B,G,TBAR0)
      MEW1=MEW0*TBAR0/TBAR
      WRITE (6,6) MEW1
6     FORMAT(1H0,5HMEW1=,F8.5)
      CALL DECAY(N1,N2,N3,MEW1,TIME,F,G)
      CALL BALANS(N1,N2,H1,EPS3,F,A,B,A1,B0,B1)
      CALL TEEBAR(N1,N2,H1,TRIPS,A,B,G,TBAR1)
2     E=ABS(TBAR-TBAR1)/TBAR
      IF (E.LE.EPS4) GO TO 3
      MEW2=((TBAR-TBAR0)*MEW1-(TBAR-TBAR1)*MEW0)/(TBAR1-TBAR0)
      WRITE (6,7) MEW2
7     FORMAT(1H0,5HMEW2=,F8.5)
      CALL DECAY(N1,N2,N3,MEW2,TIME,F,G)
      CALL BALANS(N1,N2,H1,EPS3,F,A,B,A1,B0,B1)
      CALL TEEBAR(N1,N2,H1,TRIPS,A,B,G,TBAR2)
      MEW0=MEW1
      TBAR0=TBAR1
      MEW1=MEW2
      TBAR1=TBAR2
      GO TO 2
```

Subroutine NEWMEW (continued)

```
3       MEW=MEW1
        WRITE (6,4) MEW
4       FORMAT(1H0,7HNEWMEW=,F8.5)
        RETURN
        END
```

Subroutine DECAY

```
        SUBROUTINE DECAY(N1,N2,N3,MEW,TIME,F,G)
        DIMENSION F(N1,N2),G(N1,N2),TIME(861,19),  U(19),V(19)
        REAL MEW
        EXTERNAL W0
        H3=3.141593/(N3-1)
        N0=1
        DO 200 I=1,N1
        DO 201 J=1,I
        DO 202 L=1,N3
        K=J+I*(I-1)/2
        T=TIME(K,L)
        U(L)=P(MEW,T)
        V(L)=Q(MEW,T)
202     CONTINUE
        CALL INT(N0,N3,H3,U,W0,SUMF)
        CALL INT(N0,N3,H3,V,W0,SUMG)
        F(I,J) = SUMF
        G(I,J) = SUMG
        F(J,I)=F(I,J)
        G(J,I)=G(I,J)
201     CONTINUE
200     CONTINUE
        RETURN
        END
```

Subroutine BALANS

```
        SUBROUTINE BALANS(N1,N2,H1,EPS3,F,A,B,A1,B0,B1)
        DIMENSION F(N1,N2),A1(N1),A(N1),B0(N2),B1(N2),B(N2)
        DO 230 J2=1,N2
        B0(J2)=1.0
230     CONTINUE
233     CALL BALANS1(N1,N2,H1,F,B0,A1)
        CALL BALANS2(N1,N2,H1,F,A1,B1)
        BIG=ABS(B1(1)-B0(1))/B1(1)
        DO 231 J3=2,N2
        X=ABS(B1(J3)-B0(J3))/B1(J3)
        IF (BIG.GE.X) GO TO 231
        BIG=X
231     CONTINUE
        IF (BIG.LE.EPS3) GO TO 234
        DO 232 J4=1,N2
        B0(J4)=B1(J4)
232     CONTINUE
        GO TO 233
234     R1=-H1
        DO 261 I2=1,N1
        A(I2)=A1(I2)
        R1=R1+H1
261     CONTINUE
        R2=-H1
        DO 264 J5=1,N2
        R2=R2+H1
        B(J5)=B1(J5)
264     CONTINUE
        RETURN
        END
```

Subroutine BALANS1

```
      SUBROUTINE BALANS1(N1,N2,H1,F,B,A)
      DIMENSION F(N1,N2),A(N1),B(N2),U(41)
      EXTERNAL DESTR
      N0=1
      DO 210 I=1,N1
      DO 211 J=1,N2
      U(J)=B(J)*F(I,J)
211   CONTINUE
      CALL INT(N0,N2,H1,U,DESTR,S)
      A(I)=1.0/S
210   CONTINUE
      RETURN
      END
```

Subroutine BALANS2

```
      SUBROUTINE BALANS2(N1,N2,H1,F,A,B)
      DIMENSION F(N1,N2),A(N1),B(N2),U(41)
      EXTERNAL ORIGR
      N0=1
      DO 220 J=1,N2
      DO 221 I=1,N1
      U(I)=A(I)*F(I,J)
221   CONTINUE
      CALL INT(N0,N1,H1,U,ORIGR,S)
      IF (J.NE.1) GO TO 226
      W=S
226   B(J)=W/S
220   CONTINUE
      RETURN
      END
```

Subroutine TEEBAR

```
      SUBROUTINE TEEBAR(N1,N2,H1,TRIPS,A,B,G,TBAR)
      DIMENSION G(N1,N2),A(N1),B(N2),U1(41),U2(41)
      EXTERNAL ORIGR,DESTR
      N0=1
      DO 240 I=1,N1
      DO 241 J=1,N2
      U2(J)=A(I)*B(J)*G(I,J)/TRIPS
241   CONTINUE
      CALL INT(N0,N2,H1,U2,DESTR,Y)
      U1(I)=Y
240   CONTINUE
      CALL INT(N0,N1,H1,U1,ORIGR,TBAR)
      WRITE (6,245) TBAR
245   FORMAT(1H0,22HAVERAGE TRAVEL TIME = ,F8.4,8H MINUTES)
      RETURN
      END
```

Subroutine AVTIME

```
      SUBROUTINE AVTIME(N1,N2,H1,A,B,G,AVTIME1,AVTIME2,MEW,TRIPS)
      DIMENSION G(N1,N2),A(N1),B(N2),AVTIME1(N1),AVTIME2(N2),EMT1(41)
     1EMT2(41),U1(41),U2(41)
      REAL MEW
      EXTERNAL ORIGR,DESTR
      N0=1
      WRITE (6,306)
306   FORMAT(1H1,32H R1       AVTIME1(R1)      EMT1(R1))
      DO 300 I=1,N1
      DO 301 J=1,N2
      U2(J)=A(I)*B(J)*G(I,J)/(2.0*3.141593)
```

Subroutine AVTIME (continued)

```
301     CONTINUE
        CALL INT(N0,N2,H1,U2,DESTR,SUM1)
        AVTIME1(I)=SUM1
        EMT1(I)=EXP(MEW*AVTIME1(I))
        R1=(FLOAT(I)-1.0)*H1
        WRITE (6,305) R1,AVTIME1(I),EMT1(I)
305     FORMAT(1H0,F4.1,4X,F10.5,4X,F10.5)
300     CONTINUE
        WRITE (6,307)
307     FORMAT(1H1,32H R2       AVTIME2(R2)       EMT2(R2))
        DO 308 J1=1,N2
        DO 309 I=1,N1
        U1(I)=A(I)*B(J1)*G(I,J1)/(2.0*3.141593)
309     CONTINUE
        CALL INT(N0,N1,H1,U1,ORIGR,SUM2)
        AVTIME2(J1)=SUM2
        EMT2(J1)=EXP(MEW*AVTIME2(J1))
        R2=(FLOAT(J1)-1.0)*H1
        WRITE (6,305) R2,AVTIME2(J1),EMT2(J1)
        EMT2(J1)=1.0
308     CONTINUE
        CALL INT(N0,N1,H1,AVTIME1,ORIGR,S0)
        CALL INT(N0,N2,H1,AVTIME2,DESTR,SD)
        CALL INT(N0,N2,H1,EMT2,DESTR,TPD)
        S=2.0*3.141593/TRIPS
        TO=S0*S
        TD=SD*S
        DTRIPS=2.0*3.141593*TPD
        WRITE (6,313) TO,TD,DTRIPS
313     FORMAT(1H0,3HTO=,F10.5,5X,3HTD=,F10.5,7HDTRIPS=,F10.1)
        RETURN
        END
```

Subroutine ODEN

```
        SUBROUTINE ODEN(N1,N2,H1,F,A,B,0)
        DIMENSION F(N1,N2),A(N1),B(N2),0(N1),U1(41),U2(41)
        EXTERNAL ORIGR,DESTR
        WRITE (6,618)
618     FORMAT(1H1,12HORIGINS TEST)
        N0=1
        R1=-H1
        DO 610 I=1,N1
        R1=R1+H1
        IF (I.GT.1) GO TO 616
        0(I)=0.0
        S2=0.0
        GO TO 617
616     DO 611 J=1,N2
        U2(J)=B(J)*F(I,J)
611     CONTINUE
        CALL INT(N0,N2,H1,U2,DESTR,S)
        0(I)=S*A(I)*ORIGR(R1)/R1
        S2=ORIGR(R1)/R1
617     PRINT 615, I,0(I),R1,S2
615     FORMAT(1H0,2HA(,I2,4H) = ,F10.3,5X,2HO(,F4.1,4H) = ,F10.3)
610     CONTINUE
        WRITE (6,620)
620     FORMAT(1H0,9HDEST TEST)
        R2=0.0
        DO 621 J=2,N2
        R2=R2+H1
        DO 622 I=1,N1
        0(I)=A(I)*F(I,J)
```

Subroutine ODEN (continued)

```
622    CONTINUE
       CALL INT(N0,N1,H1,O,ORIGR,S)
       S3=S*B(J)*DESTR(R2)/R2
       S4=DESTR(R2)/R2
       PRINT 623,J,S3,R2,S4
623    FORMAT(1H0,2HB(,I2,4H) =  ,F10.3,5X,2HD(,F4.1,4H) =  ,F10.3)
621    CONTINUE
       RETURN
       END
```

Subroutine KIRBY

```
       SUBROUTINE KIRBY(N1,N2,H1,TRIPS,A,B,ASTAR,BSTAR,F)
       DIMENSION A(N1),B(N2),ASTAR(N1),BSTAR(N2),U1(41),AS(41),BS(4)
      10(41)
       EXTERNAL ORIGR,DESTR
       N0=1
       CALL INT(N0,N2,H1,B,DESTR,S1)
       DO 821 I1=1,N1
       U1(I1)=A(I1)*S1
821    CONTINUE
       CALL INT(N0,N1,H1,U1,ORIGR,S2)
       GAMMA=S2*4.0*3.141593**2/TRIPS
       CALL INT(N0,N1,H1,A,ORIGR,S3)
       GAMMA1=S3*2.0*3.141593/TRIPS
       CALL INT(N0,N2,H1,B,DESTR,S4)
       GAMMA2=S4*2.0*3.141593/TRIPS
       W=SQRT(GAMMA*TRIPS)
       WRITE(6,822) GAMMA,GAMMA1,GAMMA2
822    FORMAT(1H1,F10.4,5X,F10.4,5X,F10.4)
       R1=-H1
       DO 825 I2=1,N1
       R1=R1+H1
       ASTAR(I2)=A(I2)*GAMMA/GAMMA1
       AS(I2)=ASTAR(I2)/W
       WRITE (6,827) R1,ASTAR(I2)
827    FORMAT(1H0,6HASTAR(,F4.1,2H)=,F10.4)
825    CONTINUE
       R2=-H1
       DO 826 J2=1,N2
       R2=R2+H1
       BSTAR(J2)=B(J2)*GAMMA/GAMMA2
       BS(J2)=BSTAR(J2)/W
       WRITE (6,828) R2,BSTAR(J2)
828    FORMAT(1H0,6HBSTAR(,F4.1,2H)=,F10.4)
826    CONTINUE
       CALL ODEN(N1,N2,H1,F,AS,BS,O)
       RETURN
       END
```

Subroutine TPLOT

```
        SUBROUTINE TPLOT(N1,N2,N3,H1,MEW,A,B,TIME,TSTAR,T)
        DIMENSION A(N1),B(N2),TSTAR(60),TIME(861,19),  T(60)
        REAL MEW
        H3=3.141593/(N3-1)
        DO 101 L=1,60
        TSTAR(L)=0.0
        T(L)=0.0
101     CONTINUE
        R1=-H1
        DO 102 I=1,N1
        R1=R1+H1
        R2=-H1
        DO 103 J=1,N2
        R2=R2+H1
        DO104 M=1,N3
        IF (J.GT.I) GO TO 110
        K=J+I*(I-1)/2
        GO TO 111
110     K=I+J*(J-1)/2
111     T=TIME(K,M)
        ST=A(I)*B(J)*ORIGR(R1)*DESTR(R2)
        STAR=ST*EXP(-MEW*T)*4.0*H1*H1*H3*3.141593
        IF (M.EQ.1.OR.M.EQ.N3) STAR=STAR/2.0
        BAR=T*STAR
        T0=0.0
        DO 105 N=1,60
        T0=T0+1.0
        IF (T.GT.T0) GO TO 105
        TSTAR(N)=TSTAR(N)+STAR
        T(N)=T(N)+BAR
        GO TO 104
105     CONTINUE
104     CONTINUE
103     CONTINUE
102     CONTINUE
        WRITE (6,107)
107     FORMAT(1H1,2HT1,5X,2HT2,5X,12HNO. OF TRIPS)
        TS=0.0
        TRPS=0.0
        DO 106 L1=1,60
        TS=TS+T(L1)
        TRPS=TRPS+TSTAR(L1)
        L2=L1-1
        WRITE (6,108) L2,L1,TSTAR(L1)
108     FORMAT(1H0,I2,5X,I2,5X,F10.2)
106     CONTINUE
        TBAR=TS/TRPS
        WRITE (6,109) TRPS,TBAR
109     FORMAT(1H0,6HTRIPS=,F10.2,5X,5HTBAR=,F7.3)
        RETURN
        END
```

Subroutine CROSS

```
       SUBROUTINE CROSS(N1,N2,N3,N4,H1,H3,MEW,TIME,F,A,B,C,X)
       DIMENSION A(N1),B(N2),F(N1,N2),U1(41),U2(41),U3(21),CI(21),CO(21),
      1DC(21),C(N4),X(N4),TIME(861,19),  IN(21),OUT(21),ALL(21),U22(41)
       REAL MEW,IN
       EXTERNAL ORIGR,DESTR,W0
       WRITE (6,800)
800    FORMAT(1H1,124H R          CI(R)           CO(R)           DC(R)
      1    C(R)            X(R)            XI(R)          XO(R)          X
      2D(R)    )
       N0=1
       N4=(N1+1)/2
       W=2.0*3.141593
       X0=0.0
       DO 801 M=2,N4
       N=2*M-1
       K=2*M-1
       IF (K.GE.N2) GO TO 815
       GO TO 816
815    K=N2
816    DO 802 J1=1,N2
       DO 803 I1=1,N1
       U1(I1)=A(I1)*F(I1,J1)
803    CONTINUE
       CALL INT(N0,N,H1,U1,ORIGR,S1)
       CALL INT(N,N1,H1,U1,ORIGR,S2)
       U2(J1)=B(J1)*S1
       U22(J1)=B(J1)*S2
802    CONTINUE
       CALL INT(K,N2,H1,U2,DESTR,S3)
       CALL INT(N0,K,H1,U2,DESTR,S4)
       CALL INT(K,N2,H1,U22,DESTR,S5)
       CALL INT(N0,K,H1,U22,DESTR,S6)
       CO(M)=S3*W
       CI(M)=S6*W
       IN(M)=S4*W
       OUT(M)=S5*W
       ALL(M)=CO(M)+CI(M)+IN(M)+OUT(M)
       R=(FLOAT(M)-1.0)*2.0*H1
       R1=R-H1
       DO 808 I3=N,N1
       R1=R1+H1
       R2=R-H1
       DO 809 J3=K,N2
       R2=R2+H1
       CALL TANGLE(R,R1,A1)
       CALL TANGLE(R,R2,A2)
       A0=A1+A2
       H4=(3.141593-A0)/(N3-1)
       AN=A0-H4
       DO 810 L=1,N3
       AN=AN+H4
       ANG=0.0
       DO 821 L3=2,N3
       ANG=ANG+H3
       IF (ANG.LE.AN) GO TO 821
       L2=L3-1
       IF (J3.GT.I3) GO TO 830
       KK=J3+I3*(I3-1)/2
       GO TO 831
830    KK=I3+J3*(J3-1)/2
831    T=((AN+H3-ANG)*(TIME(KK,L3)-TIME(KK,L2))/H3)+TIME(KK,L2)
       GO TO 820
821    CONTINUE
820    U3(L)=P(MEW,T)
810    CONTINUE
       CALL INT(N0,N3,H3,U3,W0,S3)
       U2(J3)=B(J3)*S3
```

Subroutine CROSS (continued)

```
809     CONTINUE
        CALL INT(K,N2,H1,U2,DESTR,S2)
        U1(I3)=A(I3)*S2
808     CONTINUE
        CALL INT(N,N1,H1,U1,ORIGR,S1)
        DC(M)=S1*4.0*3.141593
        C(M)=CI(M)+CO(M)+DC(M)
        X(M)=C(M)/(2.0*3.141593*R)
        XI=CI(M)/(2.0*3.141593*R)
        XO=CO(M)/(2.0*3.141593*R)
        XD=DC(M)/(2.0*3.141593*R)
        WRITE (6,807) R,CI(M),CO(M),DC(M),C(M),X(M),XI,XO,XD
807     FORMAT(1H0,F4.1,5X,F10.2,5X,F10.2,5X,F10.2,5X,F10.2,5X,F10.2,5X,F1
       10.2,5X,F10.2,5X,F10.2)
801     CONTINUE
        WRITE (6,814)
814     FORMAT(1H1,  *  R          IN          OUT          ALL
       1DLESO          CILESCO*)
        DO 812 M1=2,N4
        R=(FLOAT(M1)-1.0)*2.0*H1
        CALL INTEG(X0,R,H1,X0,X0,ORIGR,Y1)
        CALL INTEG(X0,R,H1,X0,X0,DESTR,Y2)
        DLESO=W*(Y2-Y1)
        CILESCO=CI(M1)-CO(M1)
        WRITE (6,813) R,IN(M1),OUT(M1),ALL(M1),DLESO,CILESCO
813     FORMAT(1H0,F4.1,5X,F10.2,5X,F10.2,5X,F10.2,5X,F10.2,5X,F10.2)
812     CONTINUE
        RETURN
        END
```

Subroutine TIMES

```
        SUBROUTINE TIMES(N1,N2,N3,H1,EPS1,EPS2,TIME)
        DIMENSION TIME(861,19)
        H3=3.141593/(N3-1)
        R1=-H1
        DO 11 I=1,N1
        R1=R1+H1
        R2=-H1
        DO 12 J=1,I
        R2=R2+H1
        A=-H3
        DO 13 M=1,N3
        A=A+H3
        K=J+I*(I-1)/2
        IF(I.EQ.1.OR.J.EQ.1.OR.M.EQ.1) GO TO 14
        ARAD0=3.141593/10.0
        ARAD1=3.141593-ARAD0
        IF (ARAD1.GE.A) GO TO 18
        T1=2.0*TIME(K,M-1)-TIME(K,M-2)
        GO TO 15
18      CALL TGEN(R1,R2,A,EPS1,EPS2,T1)
        GO TO 15
14      CALL TRADIAL(R1,R2,T1)
15      TIME(K,M)=T1
13      CONTINUE
12      CONTINUE
11      CONTINUE
        RETURN
        END
```

Subroutine TRADIAL

```
SUBROUTINE TRADIAL(R1,R2,T)
V1 = V(R1)
V2 = V(R2)
Z = ALOG(V2)-ALOG(V1)
X=(R2-R1+Z/0.05)/18.5
T = ABS(X)*60.0
RETURN
END
```

Subroutine TGEN

```
      SUBROUTINE TGEN(R1,R2,A,EPS1,EPS2,T)
      REAL K,KBAR
      INTEGER DOT
      CALL DOTKBAR(R1,R2,A,DOT,KBAR)
      IF (DOT.EQ.2) GO TO 51
      CALL KDIRECT(R1,R2,A,KBAR,EPS2,K)
      CALL TDIRECT(R1,R2,K,T)
      GO TO 52
51    CALL MINRAD(R1,R2,A,EPS1,EPS2,RMIN)
      CALL TTHRU(RMIN,R1,R2,T)
52    RETURN
      END
```

Subroutine DOTKBAR

```
      SUBROUTINE DOTKBAR(R1,R2,A,DOT,KBAR)
      REAL KBAR
      INTEGER DOT
      X=AMIN1(R1,R2)
      Y=AMAX1(R1,R2)
      KBAR=X/V(X)
      CALL TANGLE(X,Y,CRIT)
      IF (CRIT.GT.A) GO TO 60
      DOT=2
      GO TO 61
60    DOT=1
61    RETURN
      END
```

Subroutine KDIRECT

```
          SUBROUTINE KDIRECT(R1,R2,A,KBAR,EPS1,K)
          REAL K,K0,K1,K2,KBAR
  C       KDIRECT COMPUTES THE CHARACTERISTIC OF THE DIRECT ROUTE BETWEEN
  C       TWO POINTS WITH SPECIFIED RADII AND ANGULAR DIFFERENCE
          R0=R1*R2*SIN(A)/(SQRT(R1**2+R2**2-(2.0*R1*R2*COS(A))))
          K0=R0/V(R0)
          IF (K0.LT.KBAR) GO TO 83
          K0=KBAR*(1.0-EPS1)
  83      SPACE=KBAR*(1.0-EPS1/50.0)
          IF (K0.LT.SPACE) GO TO 88
          K0=SPACE
          GO TO 88
  88      CALL ANGLE(R1,R2,K0,A0)
          K1=K0*A/A0
          IF (K1.LT.KBAR) GO TO 82
          K1=KBAR*(1.0-EPS1/2.0)
  82      IF (K1.LT.SPACE) GO TO 89
          K1=SPACE*(1.0-EPS1/100.0)
          GO TO 89
  89      CALL ANGLE(R1,R2,K1,A1)
          ICOUNT=0
  80      E1=ABS(1.0-A1/A)
          IF (E1.LE.EPS1) GO TO 81
          E2=(ABS(1.0-A0/A1))*100.0
          IF (E2.LE.EPS1) GO TO 81
          K2=(((A-A1)*K0)-((A-A0)*K1))/(A0-A1)
          IF(K2.GT.0.0) GO TO 94
          WRITE(6,95) R1,R2,K0,K1,K2,A0,A1,A,KBAR
  95      FORMAT(1H0,5HNEGK2,2X,9F11.7)
          GO TO 81
  94      IF (K2.LT.KBAR) GO TO 85
          K2=KBAR*(1.0-EPS1/3.0)
  85      IF (K2.LT.SPACE) GO TO 87
          K2=SPACE*(1.0-EPS1/10.0)
          CALL ANGLE(R1,R2,K2,A2)
          CALL SECOND(CP)
          WRITE(6,93)R1,R2,K0,K1,K2,A0,A1,A2,A,KBAR,CP
  93      FORMAT(1H0,10HSPACE JUMP,2X,11F11.7)
          K1=K2
          GO TO 81
  87      CALL ANGLE(R1,R2,K2,A2)
          A0=A1
          K0=K1
          A1=A2
          K1=K2
          ICOUNT=ICOUNT+1
          IF(ICOUNT.LT.10) GO TO 80
          WRITE(6,96)R1,R2,K0,K1,K2,A0,A1,A2,A,KBAR
  96      FORMAT(1H0,8HCOUNTOUT,2X,10F11.7)
  81      K=K1
  90      RETURN
          END
```

Subroutine TDIRECT

```
      SUBROUTINE TDIRECT(R1,R2,K,T)
      REAL K,K1
      EXTERNAL FT
C     TDIRECT COMPUTES THE TRAVEL TIME BETWEEN TWO RADII ALONG A
C     DIRECT ROUTE HAVING CHARACTERISTIC K
      Y=0.0
      X1=AMIN1(R1,R2)
      X2=AMAX1(R1,R2)
      K1=X1/V(X1)
      E=ABS(K-K1)/K
      IF(E.GT.0.01) GO TO 10
      H=(X2-X1)/10.0
      X0=X1+H
      H=(X0-X1)/10.0
      CALL INTEG(X1,X0,H,Y,K,FT,Y)
      X1=X0
10    H=(X2-X1)/10.0
      CALL INTEG(X1,X2,H,Y,K,FT,T)
      RETURN
      END
```

Subroutine MINRAD

```
      SUBROUTINE MINRAD(R1,R2,A,EPS1,EPS2,RMIN)
C     KTHRU COMPUTES THE MINIMUM RADIUS OF THE THROUGH ROUTE BETWEEN
C     TWO POINTS WITH SPECIFIED RADII AND ANGULAR DIFFERENCE
      Y=AMIN1(R1,R2)
      R0=R1*R2*SIN(A)/(SQRT(R1**2+R2**2-(2.0*R1*R2*COS(A))))
      CALL TANGLE(R0,R1,A01)
      CALL TANGLE(R0,R2,A02)
      A0=A01+A02
      RMIN0=R0
      RMIN1=R0*A0/A
      IF (RMIN1.LE.Y) GO TO 95
      RMIN1=Y*(1.0-EPS1)
95    CALL TANGLE(RMIN1,R1,A11)
      CALL TANGLE(RMIN1,R2,A12)
      A1=A11+A12
      ICOUNT=0
90    E1=ABS(1.0-A1/A)
      IF (E1.LE.EPS1) GO TO 91
      RMIN2=(((A-A1)*RMIN0)-((A-A0)*RMIN1))/(A0-A1)
      IF (RMIN2.LE.Y) GO TO 96
      RMIN2=Y*(1.0-EPS1/2.0)
      RMIN1=RMIN2
      GO TO 91
96    CALL TANGLE(RMIN2,R1,A21)
      CALL TANGLE(RMIN2,R2,A22)
      A2=A21+A22
      A0=A1
      RMIN0=RMIN1
      A1=A2
      RMIN1=RMIN2
      ICOUNT=ICOUNT+1
      IF (ICOUNT.GE.10) GO TO 91
      GO TO 90
91    RMIN=RMIN1
      RETURN
      END
```

Subroutine TTHRU

```
      SUBROUTINE TTHRU(RMIN,R1,R2,T)
      CALL THRU(RMIN,R1,T1)
      CALL THRU(RMIN,R2,T2)
      T=T1+T2
      RETURN
      END
```

Subroutine THRU

```
      SUBROUTINE THRU(RMIN,R,T)
      REAL K
      EXTERNAL FT
C     THRU COMPUTES THE TRAVEL TIME FROM A MINIMUM RADIUS RMIN
C     THE RADIUS R
      K=RMIN/V(RMIN)
      H1=RMIN/50.0
      H2=ABS(R-RMIN)/20.0
      H=AMIN1(H1,H2)
      X=SQRT(4.0*RMIN*H+4.0*H*H)
      VMIN=V(RMIN)
      Y1=X*60.0/VMIN
      R0=RMIN+2.0*H
      H=ABS(R-R0)/20.0
      CALL INTEG(R0,R,H,Y1,K,FT,T)
      RETURN
      END
```

Subroutine ANGLE

```
      SUBROUTINE ANGLE(R1,R2,K,A)
      REAL K,K1
      EXTERNAL FA
C     ANGLE COMPUTES THE ANGULAR DIFFERENCE BETWEEN TWO POINTS OF GIVEN
C     RADII,ALONG A DIRECT ROUTE WITH SPECIFIED CHARACTERISTIC
      Y=0.0
      X1=AMIN1(R1,R2)
      X2=AMAX1(R1,R2)
      K1=X1/V(X1)
      E=ABS(K-K1)/K
      IF(E.GT.0.01) GO TO 10
      H=(X2-X1)/10.0
      X0=X1+H
      H=(X0-X1)/10.0
      IF(K1.GT.K.AND.K.GT.0.0) GO TO 11
      WRITE(6,12) R1,R2,K,K1
12    FORMAT(1H0,4F11.7)
11    CALL INTEG(X1,X0,H,Y,K,FA,Y)
      X1=X0
10    H=(X2-X1)/10.0
      CALL INTEG(X1,X2,H,Y,K,FA,A)
      RETURN
      END
```

Subroutine TANGLE

```
      SUBROUTINE TANGLE(RMIN,R,A)
      REAL K
      EXTERNAL FA
C     TANGLE COMPUTES THE CRITICAL ANGLE OF A ROUTE BETWEEN TWO RADII
      K=RMIN/V(RMIN)
      H1=RMIN/50.0
      H2=ABS(R-RMIN)/20.0
      H=AMIN1(H1,H2)
      Y=ACOS(RMIN/(RMIN+2.0*H))
      R0=RMIN+H*2.0
      H=ABS(R-R0)/20.0
      CALL INTEG(R0,R,H,Y,K,FA,A)
      RETURN
      END
```

Subroutine INTEG

```
      SUBROUTINE INTEG(X0,X1,H,Y0,K,FY,Y)
      REAL K
C     INTEG INTEGRATES THE FUNCTION FY(R,K) FROM R=X0 TO R=X1 BY
C     INCREMENTS OF H AND ADDS THE RESULT TO THE INITIAL VALUE Y0
      Y=Y0
      IF (X0.GE.X1) GO TO 165
      ICOUNT=0
      X=X0-H
160   X=X+H
      TEST=X/V(X)
      IF(TEST.GT.K) GO TO 170
      WRITE(6,171) X,TEST,K,X0,X1,Y0
171   FORMAT(1H0,8HTEST X/K,6F20.10)
170   Z=FY(X,K)
      IF(Z.GE.0.0) GO TO 172
      WRITE(6,173) X,Z,K,X0,X1,Y0
173   FORMAT(1H0,5HZ OUT,6F20.10)
      STOP
172   IF (ICOUNT-1) 161,162,163
161   Z1=Z
      GO TO 164
162   Z2=Z
      GO TO 164
163   Z3=Z
      Y1=(Z1+4.0*Z2+Z3)*H/3.0
      Y=Y+Y1
      Z1=Z3
      ICOUNT=0
164   ICOUNT=ICOUNT+1
      IF (X.GE.X1) GO TO 165
      GO TO 160
165   RETURN
      END
```

Subroutine INT

```
      SUBROUTINE INT(N0,N,H,U,W,S)
      DIMENSION U(N)
      ICOUNT=0
      S=0.0
      DO 30 I=N0,N
      R=(FLOAT(I)-1.0)*H
      Z=U(I)*W(R)
      IF (ICOUNT-1) 31,32,33
31    Z1=Z
      GO TO 30
32    Z2=Z
      GO TO 30
33    Z3=Z
      S1=(Z1+4.0*Z2+Z3)*H/3.0
      S=S+S1
      Z1=Z3
      ICOUNT=0
30    ICOUNT=ICOUNT+1
      RETURN
      END
```

Functions

```
      FUNCTION ORIGR(R1)
      ORIGR=1164.1*R1**1.982*EXP(-0.439*R1)
      RETURN
      END


      FUNCTION DESTR(R2)
      TRIPS=166000
      DESTR=4677.0*R2**0.549*EXP(-0.298*R2)*TRIPS/168700
      RETURN
      END


      FUNCTION P(MEW,T)
      REAL MEW
      P=2.0*EXP(-MEW*T)
      RETURN
      END


      FUNCTION Q(MEW,T)
      REAL MEW
      Q=4.0*EXP(-MEW*T)*T*3.141593
      RETURN
      END


      FUNCTION W0(A)
      W0=1.0
      RETURN
      END


      FUNCTION FA(R,K)
      REAL K
C     FA(R,K) IS THE INTEGRAND THAT IS USED TO COMPUTE THE ANGULAR
C     DIFFERENCE ALONG A ROUTE HAVING CHARACTERISTIC K
      VR=V(R)
      D=R*R-K*K*VR*VR
      IF(D.GT.0.0) GO TO 1
      WRITE(6,2) R,K,VR,D
2     FORMAT(1H0,6HFA OUT,4F20.10)
      FA=-1.0
      GO TO 3
1     FA=K*VR/(R*SQRT(D))
3     RETURN
      END


      FUNCTION FT(R,K)
      REAL K
C     FT(R,K) IS THE INTEGRAND THAT IS USED TO COMPUTE THE TRAVEL
C     TIME ALONG A ROUTE HAVING CHARACTERISTIC K
      V = V(R)
      FT=(R/(V*SQRT(R**2-(K**2*V**2))))*60
      RETURN
      END


      FUNCTION V(R)
C     V(R) IS THE VELOCITY FIELD
      V=18.5-12.5*EXP(-0.05*R)
      RETURN
      END
```

Reprinted from *Papers of the Regional Science Association*, **19**, 7-19, 1967

GLOBAL SCIENCE AND THE TYRANNY OF SPACE

WILLIAM WARNTZ, Harvard University

Space is a tyrant and distances enforce his rule. He militates against us, often disposing of what we propose if our plans ignore his influence. The revolution against him is already well begun, however. Among the most disloyal subjects are geographers and regional scientists. Their attack on space is premeditated, calculating, and unremitting. They aim to understand him completely, the better to channel his influences to their own ends, and they are willing to study long hours and hold frequent conferences to achieve this. They know the rule of space is not whimsical or capricious. That it is systematic and orderly has been glimpsed. When thoroughly understood, advantage will redound to mankind.

Thomas Macaulay sensed this when he wrote, "Of all the inventions, the alphabet and the printing press alone excepted, those inventions which abridge distance have done most for the civilization of the species."

Every age and every society have witnessed, whether in a learned way or not, attempts to improve the convenience, speed, comfort, economy, and reliability with which the physical distances separating people from each other and from their required physical materials are overcome.

As we continue our studies of the role of distance as a dimension of society, we are constantly reminded that its importance is to be judged not in physical units of length alone but rather in terms of cost distances, time distances, and the like. We are informed of the concept of the elastic mile. The straight-line distance over the earth's surface is not, we are told, to be regarded necessarily as the effective distance separating places. Rather, circuitous land routings between places, for example, may, within limits, prove more economical if intervening difficult terrain and the high cost of traversing it can thus be avoided. Then, too, the favorable ratio of water transport rates to land transport rates frequently occasions a willingess to add substantial amounts to the geographical distance involved, the result being significant economic gain.

To man, operational space on the earth's surface is not at any one moment to be conceived as representing a single geometric unity. For example, many conceptual surfaces may be regarded as overlying the physical surface of the earth, each such conceptual surface being defined operationally in its appropriate terms. Such surfaces coexist but do not coincide. Their reconciliations present problems of major significance to planners and modern scholars of society. One man's crooked lines are another man's geodesics. Just as one man's trivia are another man's data.

For sake of analysis, we can imagine a portion of the physical surface of the earth as transformed into a time surface or a cost surface, for example, with optimum paths regarded as geodesics on their appropriate surfaces. Rarely do the paths of time or cost geodesics correspond completely to the pure physical geographical geodesics of minimum distances between places, but they may be regarded as subject to certain influences tending to cause them to do so. We will return to this idea later.

Time geodesics or cost geodesics, although generally not straight paths over the earth's physical surface, are nevertheless straight with respect to the surfaces on which they lie. However much they may bend and turn with respect to the earth's surface as the datum, they do not turn at all but rather move only straight forward with respect to the surface in whose terms they are defined.

The general analytical approach to the study of such paths as these of economic and social significance is through differential geometry and calculus of variations, and the principal concepts employed lie within a generalized theory of refraction, a general spatial *lex parsimoniae*. Thus, for example, to establish the minimum land-acquisition-cost path between two points within a region, we assign to land values, when integrated around a given point, a role isomorphic with that of an index of refraction. And, for minimum time paths for aircraft on long transoceanic flights, the effect of the wind, too, is to be understood in terms of refraction, although here the case is not simply one of a scalar value but rather that of a vector, so that routes in opposite directions between two points do not coincide as is the case for minimum land-acquisition-cost paths.

Many other examples might be given in addition to the ones mentioned above, but the distance surfaces, time surfaces, and cost surfaces that these represent are indicative of the wide application of the basic ideas of surfaces and paths generally. This, of course, as noted before includes a part of the general theory of geodesics and of refraction. Intensive study and the *literal* application of geometrical models to the real earth and especially to the various conceptual surfaces covering it seem called for at this stage in the development and growth of our science of social and economic phenomena toward a truly unified discipline. All that has gone before is introduction. We again acknowledge the importance of the elastic mile.

However, an interesting, important, and altogether reasonable history of civilization could be written about the persistent and pervading attempts of the human race to straighten all paths, to minimize all distances, and to merge all operational geometries into one. Thus, we can suggest that "the longest revolution," the transport revolution, has as its goal the making of the earth into one, smooth and spherical in social and economic terms, thus reducing to the minimum the deterrents imposed by the physical reality of space. A physical least distance path represents the *minimum mimimorum* toward which the paths of all other geodesics, however defined, converge. This is accomplished in two ways. One way is represented in the changing of the physical environment by man. Man changes the face of the earth by bridging streams, leveling hills, improving trafficability by laying down hard surfaces, and in general removing or modifying the barriers impeding maximum accessibilities. These modifications have sometimes taken the form of major earth surgery as in the case of the Panama Canal. Completely and unmitigatedly equal accessibility in all directions around any point on a surface, however, is ultimately limited not just by natural imperfections but also by the necessity for spatially structured hierarchical patterns of transportation owing to the existence of thresholds and to limited indivisibilities relating to economies of scale. Thus, priorities and structural hierarchies of distances would exist even if the elasticity of the mile were removed. We appropriate Alonso's apt phrase and apply it in a different context, "Some wrinkles will not be ironed out."

Cost-benefits analysis is the conventional term now used to embrace, among many other things, the methods and concepts employed to determine the economic feasibility of a major project to modify the physical environment to enhance the ease and lessen the cost of transportation. Of some possible interest in this connection is an application of ideas contained within the field-quantity theory of population influence, the so-called gravity model.

Assume the intensity of the population-potential field produced by a population distributed at a given time over the area of the United States to be everywhere known. Then the definite integral of the appropriate values of potential in this geographically varying, spatially continuous field through the infinitesimal elements of population defines a value recognized as twice the total demographic energy of the spatial system, the potential at any point being unit energy. Measured and computed in convenient units, the demographic energy can be given in persons squared per mile, for example. It consists of all possible pairings in the population with each pairing's contribution to the total weighted by the reciprocal of the physical geographical distance separating its members. Such considerations are microscopic in nature. The effective distance separating the members of any pair, as noted earlier, may not at all at a given time adequately be approximated by the minimum physical distance. However, the macroscopic expression of total demograhic energy represents the one toward which the sum of all of the micropairings tends. Moreover, an extremely interesting ratio has characterized the relationship between the total demographic energy existing in the United States and the national income level at any one time. If the the national income in any one year (measured in then-current dollars) is divided by the demographic energy of the system, we have a dollar quotation of the demographic unit — proportionate to the approximate contribution to national income resulting from the exchanges between members of the population, standardized in units of two average persons a mile apart. As J. Q. Stewart has shown us, this ratio exhibited a remarkable constancy during the period, 1840-1940. Despite elastic dollars *and* elastic miles, the economic value of the demographic unit varied little from one-fourth of a cent. Business cycles were evidenced in the data, but secular trend was absent. Prior to 1840, considerable instability had occured. Following World War II, the value broke away from its earlier bonds and vaulted to the previously unreached high of one cent per demographic unit. Presumably, this resulted from the cumulative effect of leaving the gold standard, welfare state expenditures, defense expenditures, and generally enlarged federal government participation in monetary and fiscal affairs. In recent years, the value of the demographic unit has hovered about one cent, although presently its quotation is inflated to approximately one and one-half cents and appears unstable. It really matters little what the precise value is. What is of importance is that it be stable to permit such activities as insurance and banking to function efficiently. The strong suggestion here is that long-term rise in per capita income and in prices is a consequence of the number of demographic units represented by any assigned economic good or service. Incidentally, analysis of this type also allows us to make one spatial type prediction with a high degree of confidence, and that is that the average distance that nonintraurban first class letters are hauled by the postal system in the United States is 440 miles. But, let us return to the reason for introducing the concept of demographic energy.

Assume two cities on opposite sides of a mountain regarded as an impassable barrier. The effective distance separating them is not the ten-mile-long, straight line through the mountain but a necessarily much longer journey through a pass in the mountain chain. One might compute the gain in mutual demographic energy to be accomplished by tunneling the mountain. The gain in demographic energy could then be evaluated in terms of the economic value of the demographic unit. The fraction of the predicted increase in income (from shortening the distance for the two groups involved and for all others who would benefit from this restructuring) that can be recovered by tolls or taxation could be matched against expected construction and maintenance costs and the prevailing interest rate (in a time sense.) Decisions might be reached by appropriate planning officials to determine whether major earth surgery is desirable. Many times it has been. This method of analysis and synthesis suggests that a simultaneous recognition and reconciliation of what can be called interest rates, in both the temporal and the spatial sense, are not only desirable but also possible. But, a statement of the need to recognize simultaneously the role of time discounting and space discounting in an economic system is not new. In fact, words of this sort may have been the first ones uttered by Walter Isard as a child.

The reality of the joint phenomena noted above is everywhere to be seen and the pace is ever quickening. Mountains are tunneled through or even removed, valleys and pits are filled in, swamps are drained, bridges are built. The physical earth is being leveled and smoothed to an appreciable degree in many places. In addition, it is being hard-surfaced, particularly in New Jersey. From the global standpoint, it is as yet imperceptible, but it appears inevitable nevertheless.

The other way in which cost or time surface geodesics are caused to tend to converge toward physical minimum distance paths is through improvements in technology. Thus, steamships have been less influenced by wind and ocean current conditions than were sailing vessels. Modern commercial aircraft navigators in their execution of minimum time paths need to deviate less from the minimum distance tracks for transoceanic flights than did their predecessors, but, even for the present genration of jet aircaft, transatlantic flights may occasion mid-course deviation from the minimum distance path of many hundreds of miles. Even a few minutes saved by the so-called "pressure pattern" techniques are valuable in jet operations, however. If twenty-five minutes are taken as the average saving per Atlantic crossing by following the appropriate minimum time path in each case rather than the minimum distance path, the potential saving per jet crossing is $7,000. The degree of care exercised in selecting the routes and dispatching these aircraft must reflect the considerable economic risk involved.

The likely vast increases in the speed of commercial jet aircraft in the near future will undoubtedly see a decline in the significance of minimum time methods for the reason noted that increases in speed cause the minimum time path to converge toward the least distance path, and the problem will become one principally of navigating a fixed track. However, the interest in research in pressure pattern navigation will probably remain high in certain connections. For example, it now seems likely that migrating birds covering great distances do so in terms of minimum time paths in addition to all of the other remarkable navigational things that they must do. It is known that certain species migrating seasonally between the northern and southern polar regions do so along varying tracks that in general describe a vast figure of eight over the earth's surface for the round trip and seem somehow to be related to the typical pressure and hence wind fields they encounter.

As the routes for human use converge on those which we have called *minimum minimorum*, the physical oneness of the earth takes on new meaning for mankind and a real interrelatedness of all peoples becomes a distinct possibility.

The world-wide population increases, the ever quickening technological advances in transportation and communication, the vast increases in information available for all purposes, and the considerable enlargement in the number of decisions to be made daily are all evidences of increased sociological intensity and what Teilhard de Chardin has named the noösphere, that palpable though intangible tissue embracing all mankind in terms of deep human interrelationships.

Property has evaporated into something fluid and impersonal, the wealth of nations has less in common with their political boundaries than ever before, and there emerges on the scene a small but growing group of visionaries, with world horizons, willing to consider political systems more inclusive and more thoroughly global than any now in vogue.

Von Böventer has noted, "the restraining effects of transportation costs on economic exchange, in particular over long distances, have been declining. Partly as a result of modern highway construction and the cheapened cost of air travel, partly by reflecting the relative rise in value added per unit weight, the ratio of transportation costs to the price of the finished commodity has gone down. This means that more goods than ever before can be transported economically over long distances. At the same time, the mobility of both labor and capital has increased. Thus *distance has lost part of its effect as a restraint on production*."

The significance of any one unit of distance input is declining, but the total of distance inputs utilized in the world is undoubtedly increasing so rapidly that the per capita value increases as well despite the population explosion.

Thus, the whole integrated space of the earth's surface stands as a truly viable one, and continuing adjustments in the utilization of that space are to be expected. Part of man's evolution includes his adaptation to the space available. In every era, man has felt himself to be at a turning point in his ordering of his spatial activities. This is true, for evolution is continuous. But, the rate is increasing with time, so, increasingly, we seem aware of crises.

If the significance of the single unit of distance input is declining, and the total utilization of distance inputs is increasing, then the nature of the total space of all distance inputs takes on greater significance.

The earth's surface may, to a first approximation, be regarded as spherical in geometrical terms, and, in topological terms, it may be considered as a single continuous closed surface.

In physical geography, the sphericity of the earth has been recognized and assigned a role. Spatial prediction on this basis has been so very successful as to have become commonplace and no longer regarded to be spatial prediction as such. Thus, the statements about sunrise and sunset, length of day, seasonality, celestial appearances and navigation, and the like which depend in part upon the recognition of sphericity of the earth's surface are regarded as certainties and the predictiveness involved is lost sight of.

The topological nature of the surface has not received as much attention, and the requirements that this places upon the theories of spatial process in the physical geograhy of global patterns of circulation have not been recognized, to the disadvantage of those theories.

Perhaps the time of "community earth" is not too far removed into the future for us to conjecture this very day upon the meaning of a spherical closed surface for fully spatially integrated human activity upon the earth's surface. Perhaps we should not be concerned with the earliness of the hour but rather with the lateness.

If the examples we are to give mean anything, the development of a proper global science for human activities which recognizes space explicitly cannot safely be regarded merely as an extension of existing regional science efforts. In virtually every case, we have posited geometrically a flat open plane as our reference surface. The scale has been regional and the topology considered such that any completed simple boundary curve circuit on the open surface separates the area into an included one and an excluded one.

On any closed surface that is the topological equivalent of a sphere, a completed simple boundary curve circuit again divides the surface area into two parts, neither of which, however, is included or excluded. For convenience, we may regard any very small circle on the surface of a sphere, for example, as bounding a certain included area. But, the arbitrariness of this on the earth's surface is shown if we imagine a circular prison wall with an initially very small radius which is allowed to increase gradually. At first, the length of the prison wall is well approximated (but only temporarily) by the expression $2\pi r$ and the prisoner *knows* he is imprisoned. Ultimately, however, there would arrive the time when the expanding radius will have brought the entire surface of the earth minus one point behind the prison wall, and only one point, the antipodal point of the circle's center, will remain "free." The length of the prison wall will have gone from near zero to near zero through a maximum of approximatly 25,000 miles as the radius increased monotonically from near zero to approximately 12,500 miles, and that value which plays the role similar to that of pi on the plane is not a constant for the spherical surface. Although in the geopolitical realm an iron curtain may be regarded as a real phenomenon in that it restricts interaction between groups of peoples, it must be recognized that both groups are behind it or neither is.

The foregoing example is perhaps an obvious one, and it seems likely that any others we could offer at this time would be equally naive and unrewarding. We take that risk, however.

The above example, nevertheless, successfully draws attention to the basic facts that spherical geometry and not plane geometry applies to the base level on which paths are refracted and areas transformed and that the topologies of the open plane surface and of the closed spherical surface are significantly different.

Imagine now two spatially continuous distributions covering the surface of the earth. In physical geography, we might consider temperatue and pressure distributions. In socio-economic terms, the population potential field and the income potential field serve as examples. Given only that there are two continuous distributions, A and B, over the earth's surface, there will always be, as Steinhaus has noted, at least one pair of antipodal points having both the same value for A and the same value for B. Through time, the spatial distributions involved may change, and the given antipodal points initially so related as above may be no longer. However, another such pair will then exist, and there will have been a shift in position. This is the consequence of the theorem proposed by Ulam and proved by Borsuk that if a sphere is folded and distorted (but not torn) so as to be made to lie flat, there is necessarily one pair of antipodal points from the sphere which come to lie upon each other in the new situation.

In addition, Brouwer has shown that, for the earth as a whole, no assignment can be made of a circulation pattern that lacks at least one point where there is not a definite direction for the flow. At least one whirl point always must exist on the surface.

We have often noted the efficiency of the hexagonal tesselation of the plane regarding regionalization in terms of minimizing distance variances and with respect to central place theory. Now we caution that for distributions covering the entire surface of the spherical earth, a uniform hexagonal tesselation cannot be accomplished to exhaust the surface without gaps or overlaps.

The problem of the regular division of the surface of the sphere can be approached through the five regular polyhedra, the so-called Platonic bodies and the contacts they make with a circumscribed sphere. We note here that a dodecahedron yields a satisfactry pentagonal exhaustion of the sphere's surface. We leave the question of the hierarchical orderings of the space of a closed spherical surface to future global scientists with only the additional comments that an equilateral triangular tesselation is possible by way of the icosahedron. The centers of the faces of the dodecahedron form the vertices of the icosahedron, and the centers of the icosahedrons form the vertices of the dodecahedron.

As another example, we call attention to the theorem that the number of lines that can be drawn from the center of the earth to meet its actual physical surface perpendicularly is, of necessity, an even number. For a line from the center of the earth to meet its surface perpendicularly, the instantaneous gradient on the surface at that point must be zero. We have, in an earlier paper presented to this association, observed that the points meeting such a requirement consist of peaks, pits, passes, and pales. We may regard any conceptual surface (whether in physical or human geography) which is continuous over the earth's surface in these terms. Thus, we see an additional constraint to be recognized in our global science. To this, we add the knowledge that whether on a plane or a sphere, within one simple closed isoline, the number of peaks plus pits minus the number of passes plus pales must always equal one. But, since on a sphere one simple closed isoline completely bounds two areas, the number of peaks plus pits minus the number of passes plus pales therefore equals two. This is equally true of the contoured physical surface, the potential of population surface, the income potential surface, the distribution of surface pressure, the distribution of temperature, the aggregate travel distance map, and a specially conceived population density surface with all cities and towns as density peaks. Moreover, this rule remains true no matter how these distributions may change through time.

Here, however, we must add a word of caution concerning the computations required to transform discrete distribution on the sphere into spatially continuous field quantity values. As we know, in some cases in physical sciences, macroscopic field quantity values can be obtained by direct observation and measurement. The microscopic level is inferred, and concepts there are made operational by suitable computations. In social science, our direct observations and measurements are at the microscopic level. But, we find it useful, indeed necessary, to infer, to define, and to use macroscopic concepts. We make these operational by suitable computation. We note now, however, that definitions suitable for the open plane surface lead to difficulty for a closed spherical surface.

A particular case in point is the global population potential surface. On a sphere, distances can be measured on either of two arcs of the great circle on which two points lie. Moreover, if the two points are antipodal, there are infinite number of great circles on the surface passing through the two points. Thus, at the antipodal point of any unit of population, the potential contributed by that unit could be deemed to be infinitely large. On the world maps produced to date, as in *Macrogeography and Income Fronts*, we have adopted the compromise of taking only one distance, the minimum one, over the surface between two points. But, we know that the inverse distance formula for potential is good in three dimensions and that we can conceive of isopotential surfaces as well as lines. We know also that if two surfaces intersect, they do so along a line. Therefore, a map of isopotential contours for, say, the conterminous United States could be regarded as a map showing lines caused by the intersection of selected isopotential surfaces with the physical surface of the conterminous United States. Over distances as short as those in the conterminous United States, even the extreme one, the complementary great circle arcs connecting places may be disregarded safely, and the surface assumed to be a plane with the population distributed in two dimensions.

For the whole earth, however, we suggest that distances be measured not as arcs of great circles but as straight-line chords directly from point to point. We, therefore, assume that the world's population is distributed in three dimensions and that the least distance path between points which are antipodal, in a spherical sense, passes through the center of the earth. Thus, we can conceive and compute isopotential surfaces in three dimensions independent of the earth's surface. A conventional looking map of isopotential lines on the earth's surface is produced then by putting the earth into its proper position within the family of isopotential surfaces and demarcating the positions of lines produced by the intersection of the earth's surface and selected isopotential surfaces.

But, what of the center of the earth—that very, very remote place, 4,000 miles away, which man is likely to reach only after having conquered many many millions of miles to points in outer space, as for example the 49,000,000 miles to Mars when that planet and earth are closest.

Let us suppose a time when the economically effective demographic energy has increased greatly on the earth's surface, when the existing technological and financial situations have been such that the earth has been nearly smoothed and levelled, and when, at long last in human history, the earth has become meaningfully spherical. Virtually all time and cost geodesics will then have evolved into simple physical distance geodesics. (Will we by then have solved the problems of circuitous routings and frustrating delays attending surface travel within metropolitan areas?) Might not then the actual physical mass of the spherical earth itself be regarded as a barrier to travel, along with anticipatory recognition that the true shortest distance line between points lies within the earth and that such paths are along chords through the sphere. Might not the first such chord tunnel be constructed between two distant populations sharing the largest effective mutual demographic energy?

Imagine now that certain engineering problems could be overcome and that the earth could be crisscrossed with subterranean tunnels. Not only would such an arrangement reduce to an absolute minimum the distances between selected places but it would also come provided with free physical energy from the earth's own gravitational field to power the transportation system itself.

Ideas associated with such a transportation system have long been speculated upon. In 1962, such possibilities were independently recognized by the Endicott House Conference for Social Sciences and the Humanities. More recently the mathematician, Paul Cooper, has solved the equations to describe systems for earth, the moon, and several planets. Also he has investigated additional possibilities for certain curved tunnels, offering even superior time advantages.

In essence, the straight-line chord routes provide free fall through the earth between their termini on the earth's surface. By dropping into airless, frictionless tunnels in a perfectly spherical earth of uniform density with no Coriolis force induced by the earth's rotation, a vehicle would begin to fall with maximum acceleration and zero velocity. As it moved to the mid-point of its route (that point where the tunnel is closest to the earth's center and where the line from the earth's center to the tunnel meets it perpendicularly), the velocity would increase and the acceleration decrease. At this mid-point, the acceleration would be zero, but the velocity would be at a maximum. Beyond this point, the negative acceleration would cause velocity to decline, but the kinetic energy gained during the period of positive acceleration would suffice to permit the vehicle to coast "up" to the other side where it would reach its destination just as its velocity became zero. If not secured, the vehicle would oscillate back and forth forever from one terminus to the other with its velocity at any time always proportional to its distance from the mid-point of its tube. Interestingly enough, regardless of the lengths of chord tunnels, the one-way trip in any one of them would require 42.2 minutes, a round trip then of 84.4 minutes. This latter time is precisely the one published by Isaac Newton for a cannon ball fired parallel to the earth at its surface to orbit the earth, forever falling toward the earth but going fast enough so as always to miss the earth as it fell.

There have been frequent references in literature to the idea of such a subterranean tunnel transportation system. It was noted in Lewis Carroll's novel, *Sylvie and Bruno* (1893). In Clement Fezandie's novel, *Through the Earth* (1898), the vehicle in such a tunnel became stuck because Coriolis force had not been properly reckoned. A European contemporary of Fezandie, the Russian writer, Alexander Rodnykh, wrote an entire novel on the subject *Subterranean Self-Propelled Railroad between St. Petersberg and Moscow*. Other works containing the idea that come to mind are *The Tunnel* (1915) by Bernhard Kellerman, *Earth Tube* (1929) by Gawain Edwards (the nom de plume of G. Edward Pendray, rocket expert), and *Tik-Tok of Oz* by L. Frank Baum. We might add here that such a transportation system might provide a journey through Tolkien's "Middle Earth."

Let us call attention again to the fact that the one-way journeys in all chord tunnels would be equal in time regardless of the length. Travellers from everywhere to a given place would arrive at the same time regardless of variation in the lengths of their trips provided only that they all departed at the same instant. Every place on earth then would be "forty-two minutes from Broadway."

Were such a world to exist, our science would indeed need to be global, but the tyranny of distance, although not overcome, would at least be reduced to a constant. (It should be noted here, however, that certain curved tunnels could reduce all travel times below 42.2 minutes, but the times would vary.)

Let me return from the world of fantasy for some concluding remarks.

There is an ever growing danger that events will come to control men completely. If men of good will and high purpose are to control events, they must be trained to perceive and taught to act in accord with a lawful universe — social as well as physical — and their sympathies must be supplemented by the kind of knowledge that can come only from a well integrated, pertinent social science.

And, if this kind of social science is to emerge, it must be not only unified and related to all the disciplines but must be global in its pertinence. Moreover, we must recognize that only to the extent that it is a natural science of society does it stand a chance to succeed. Our special plea is that it be a science closely linked to the nature of the very earth itself including its spatial prop. erties. Global patterns are always more than the sum of regional patterns. To think otherwise is to invite the fallacy of composition.

No special alarm need be sounded to alert this audience to the certainty that mathematical expression in both its quantitative and nonquantitative domains is as much the language of social science as it is of physical science, although Francis Schiller would have us remember that since mathematics is a tool, it will, like any other tool designed by the human species, become obsolete.

Social science and physical science are but mutually related isomorphic examples of one generalized logic. In both branches, many and diverse academic specialties can be recognized usefully in terms of content, but when patterns and relationships are investigated in terms of basic categories then the true unity of all knowledge is revealed. The more we learn of any pattern, the more we learn how much it is like some other. As the most thoroughly organized and profoundly abstracted science, physics offers the best developed models for the beginning of the search for unifying principles for all knowledge. This is not to infer a dictatorship of physics or to suggest its slavish imitation by social scientists. This leads to stultification. This is rather to acknowledge the superior intellectual achievements to date of the subject. We predict with confidence, however, that more thorough investigation of sociological phenomena will produce models that can be reflected back into physics and improve analysis there, too. General systems theory and energy analysis strongly suggest this.

In physics, the profound abstraction known as energy has supplemented space, time, and matter as an instrument of thought, and, even more importantly, energy is the concept which has integrated the diverse findings of many specialists in research.

In social science, research has led to the recognition of "social energies." Whether with physical or social phenomena, energy is reckoned as the product of an intension by its appropriate extension. The repeated appearance and reliability of this one pattern provide a clue to the needed high generality and a way of demonstrating the true unity of all knowlede in the manner of Leibniz, Wiener, and Stewart whose philosophies center about concepts of universal symbolism and a calculus reasoning.

Whether it is this plan or some more profound and recondite rival that comes to provide the needed organization for the social sciences, the fact remains that an excruciating urgency exists.

Military engineers threaten to turn our planet into a temporary star. Yet, too few people seem aware that "humanity is on the edge of destruction for lack of social concepts to match its physical powers."

The recent and continuing great increases in human population and the improved means for the flow of information and communication among them seem not to be easing the world's tension but rather to offer increasing opportunities for mischief, a potential which unfortunately currently seems nearly fully realized.

Since the middle of the seventeenth century the world's population has grown from five hundred million to three billion persons, a six-fold increase. The world's population doubled between 1650 and 1850, growing beyond the one billion mark. It doubled again by 1930 to reach two billion in only eighty years. And since then, the rate of growth has accelerated steadily. At present, over fifty million persons are added to the world's population annually. It has taken all of the vast ages of history to build the world's present population of three billions, and, yet, this is likely to be doubled again before the presnt century is done! And with this very likely will be an attendant increase many times over in social problems.

The last three hundred years have brought mankind to the modern age through a series of agricultural, industrial, communication, transportation, and other technological revolutions. These developments have made possible the support of mammoth populations in numerous areas of the world. However, many of the technological advances are only beginning to touch the less developed areas where living levels for two-thirds of the world's people are only a little, if any, above what they were during much of the earlier history of the race, but with contacts with the rest of the world that are greatly increased.

But as enormous technologically as may be the task of procuring adequate food, clothing, and shelter for all of the world's people, that problem pales in comparison with the sociological, political, and psychological strains, stresses, and tensions that beset all of mankind and especially those of the prosperous countries.

Although the majority in our own nation may feel that security and solace are to be found in winning a purely technological space race against an able adversary or in a military adventure and express a willingness to having many of our national resources allocated to carrying out these undertakings, I am sure that there must be a distinct though smaller group equally concerned lest proper attention be denied to the social sciences and humanities, the proper development of which ultimately will point the true road to human well-being.

Without intellectuals guiding public sympathy, without a high level of competence in the relevant social sciences, and without a general stirring of intellectual life at all levels, the most daring, powerful, and pertinent new ideas may die a-borning.

Research in these fields is one of our urgent concerns today because history shows us repeatedly and dramatically the truth of Lord Bacon's dictum, "Knowledge is power."

If the public mind can ever be educated fully to realize that the spearhead of true progress is philosophical imagination and pure rationality complementing empirical observation, man might realize the true essence of humanity, and no practicality could deter us from that next neceessary revolution in science, that next necessary great advance of knowledge, a truly natural science of society, giving proper emphasis to spatial as well as nonspatial processes and energy flows.   And could not this advance be spearheaded by regional scien-tists and geographers?

A note on surfaces and paths and applications to geographical problems

by William Warntz

Recent gains in the development of general geography as a science at the theoretical–predictive level have been acomplished largely, perhaps, as a result of the simultaneous increase in both the sophistication and the naiveté with which geographers view the phenomena of the real world. This seeming paradox is explainable, for example, with reference to mathematics. More than ever before, geographers are using the tools of calculus, probability, topology, symbolic logic, the various algebras, geometries, for example, are being taken more literally than ever before.

We are aware of the topological nature of problems in non-spatial sciences like chemistry involving connectives or bondings of molecules, and of those concerning hierarchical chains of command, responsibility, and interactions as recognized in certain of the non-spatial social sciences. In addition, in general classificatory science and in the mathematical set theory underlying it, use is made of Venn diagrams, which utilize topological properties of an idealized space to portray graphically such relations as are implied in subsets, intersects, and unions, etc. But, we can take Venn diagrams in a far more literal sense than they were originally intended and by substituting real space and attendant phenomena for ideal space and by insisting upon utilization of all of the geometric properties involved as well as just the topological ones, geographers can reinterpret, add to, and refine the conventional concepts in the methodology of uniform regional geography and provide it with a basis in logic. Further discussion of this important topic lies outside the present paper and is reserved for presentation elsewhere. Of course, geographers also take the spatial considerations in topology literally in analysis of highway networks, river systems, and so on.

With regard to geometries, one also can cite numerous examples of geometrical solutions to problems which are not inherently spatial or in which the problems have been abstracted from space, and geometry is employed *only* by analogy. Included would be such things as one approach in economics to consumer tastes, prices, and preferences. Indifference curves are used to portray a surface of satisfaction. Various paths on this surface have meaning with regard to the income effect and the substitution effect. Other examples abound in economics.

In chemistry the use of surfaces and paths to show relationships in non-spatial thermodynamics is due to the nineteenth century American scientist, Josiah Willard Gibbs. His methods of geometrical representation of thermodynamic properties of substances by means of surfaces showed, for example, how to diagram water as it undergoes changes from solid to liquid to gas. So impressive was this work that the brilliant British scientist, Clerk Maxwell, saluted Gibbs by building for him a plaster model entitled, "a statue of water." In general the mathematics of response surfaces, supported by appropriate statistical measures now appear as well in many non-spatial sciences, e.g. psychology, learning theory, and so on.

Today geographers are taking the *geo* in geometry literally and the study of earth related surfaces and paths has now been expanded far beyond its original application to such things as land forms, contour mapping, drainage patterns, temperatures, pressures, precipitation, and the like in physical geography alone.

The modern geographer conceives of surfaces based also on social, economic, and cultural phenomena portraying not only conventional densities but other things such as field quantity potentials, probabilities, costs, times, and so on. Always, however, these conceptual surfaces may be regarded as overlying the surface of the real earth and the geometric and topological characteristics of these surfaces, as transformed, thus describe aspects of the geography of the real world.

Elaboration of additional important ideas may be found in William Bunge's *Theoretical Geography*, Lund, Sweden, 1962. Bunge has pointed out the necessity and the efficiency of recognizing the inseparability of *geometry* as the mathematics, i.e., the language, of spatial relations and *geography*, as the science of spatial relations.

Illustrations of the application of the ideas, the geometry of surfaces and paths to a number of geographical problems can readily be given. The examples following have been selected because of their non-spatial diversity.

Our first example concerns any conformal map projection with the one exhibited here in figure 1, the well known Mercator projection—equatorial case. An exact mathematical isomorphism exists between the paths of light rays in an isotropic medium with an index of refraction varying from point to point but constant in all directions around any given point and the least distance paths or great circle arcs on the earth's spherical surface as represented on a conformal map. Let f be the scale of a conformal map at any point—expressed as the fractional ratio of distance on the map to actual distance on the earth, and let ds be the infinitesimal line element measured on the map. Then, the value $\int \frac{1}{f} ds$ is a minimum for the great circle track between any two given points as compared with the values obtained by integrating along any nearby alternative paths. This can be stated as a calculus of variations problem.

The designation of great circle tracks can also be accomplished by graphical portrayal of a surface and a gradient path. Shown in figure 1 is a Mercator map with a distance surface based on London depicted by isodistance lines in statute miles with a constant interval. On the real earth such lines would be concentric circles at first with increasing circumferences and then decreasing to zero at the antipodal point. When portrayed on the Mercator projection these circles are transformed, for although the Mercator is conformal in the small, it cannot show correct shapes in the large because of scale change. On a spherical earth great circles from all points to a given point would be orthogonal trajectories to the iso-distance circles centered on the given point. On this conformal map that property is retained but the great circles must bend to achieve it. Notice then, the presentation in figure 1 of the isolines of the distance surface and the orthogonal trajectories or gradient paths providing one solution to the problem of great circle determination. (Note that an actual geodesic on the map plane, the straight line, traces out a rhumb or constant heading line on the earth.)

Figure 1 A shows another conformal map—a stereographic with the center of projection, in this case, at the South Pole. Shown on this map are certain iso-distance circles in statute miles and great circle paths based on Salisbury, Southern Rhodesia. Again the gradient path for great circles on the distance surface is found. The rule of orthogonal trajectory applies to this as to all other conformal maps.

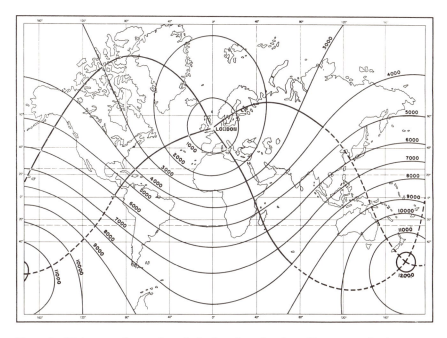

Figure 1. Distance surface and geodesics based on London. Mercator projection.

The stereographic has a number of interesting properties. All great or small circles, or arcs of circles, on the earth's surface map as circles or arcs of circles on the stereographic. But concentric circles on the earth, while mapping as circles, do not, in general map as concentric circles. The exception, of course, is the case for the center of the projection. In figure 1 A (assuming a perfectly spherical earth) small circles of latitude (these may be regarded as iso-distance lines on the earth) do map as concentric circles about the South Pole. However, let us look at the iso-distance circles about Salisbury. These are concentric circles on a spherical earth but on the plane stereographic projection they map as non-concentric circles—or non-concentric arcs of circles since the entire earth is not shown on one stereographic projection.

The centers of these mapped iso-distance circles about Salisbury march steadily away from Salisbury along the straight line on the plane of the projection from the center of the projection (the South Pole) extended through Salisbury. The direction of movement, of course, is away from the center of the projection. The iso-distance circle that passes through the antipodal point of the center of the projection will map as a straight line. For this map the antipodal point of the center of the projection is the North Pole. (This is also the center of perspective for this

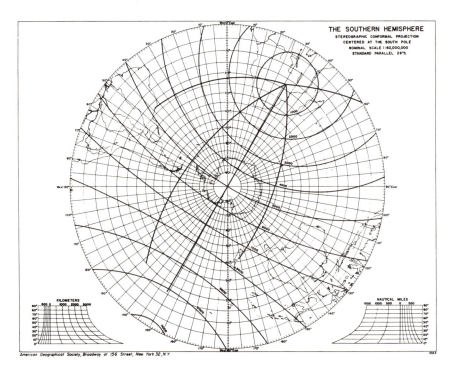

Figure 1A. Distance surface and geodesics based on Salisbury, South Rhodesia.

projection.) So, the iso-distance circle about Salisbury that passes through the North Pole maps as a straight line and its center on the plane of projection lies at infinity, in fact at both plus and minus infinity if we adopt that convention. The next iso-distance circle out from Salisbury will map with negative curvature and its center on the plane of the projection will lie between the point at negative infinity and the antipodal point of Salisbury. The centers of the succeeding iso-distance circles based on Salisbury will march ever closer to Salisbury's antipodal point. This path, of course, is the straight line on the plane from the point of minus infinity to the antipodal point of Salisbury and which when extended would also pass through the South Pole (and beyond through Salisbury and to the point of plus infinity noted above).

Once the positions of the selected iso-distance circles have been determined the drawing of the least earth distance path or geodesic from any point to Salisbury is a simple matter. It is the orthogonal trajectory or gradient path on this conformal map. All places are relative sources for Salisbury as the single sink. The refraction analogy holds again and the reciprocal of the map scale at any point may be regarded as an index of refraction for the geodesic there.

It is interesting to note that the family of geodesics passing through Salisbury which, of course, are great circles on the earth's surface also map as circles on the stereographic.

On the surface of the spherical earth, the latitude and longitude circles may be regarded, respectively, as distance circles about, and geodesics passing through, the geographic poles. Thus, our Salisbury case resembles the polar case on any non-polar stereographic.

The next example considers the problem of determining the path for minimum time enroute for an aircraft between two airports separated by a broad expanse with winds of varying direction and velocity. Almost never will the least distance path, i.e., great circle route, afford the minimum time enroute. Generally, it is possible to deviate from this great circle route to enhance the speed over the earth's surface by utilizing more favorable winds. This will be attempted, ordinarily, so long as the addition to speed is proportionally greater than the addition to distance. An elegant mathematical solution to this calculus of variations problem exists which, however, is not practical. Essentially the problem is one of the path which minimizes the value, $\int \frac{1}{v} ds$ when v is the speed of the aircraft over the earth's surface and ds is as above. Difficulty arises from the fact that velocity and direction of the aircraft over the surface depends upon the vector of heading and true air speed of the aircraft combined with that of wind direction and velocity. The resulting tail wind or head wind component at any given point in the wind field is not independent of the aircraft's actual heading. But, a graphical solution can be used to develop

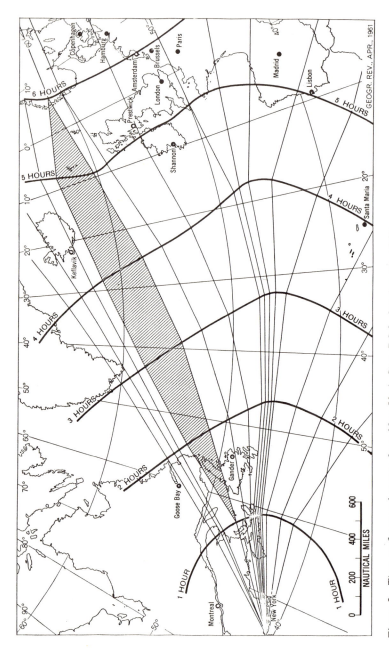

Figure 2. Time surfaces and routes from New York City, DC-8, October 17, 1960.

a family of isotime lines and one also of course lines based on a given departure point and showing the times and routes for minimum time flights to all destinations. The method involved is that modified by including wind vector considerations from the seventeenth century Dutch scientist, Huygens, relating to sound and light refraction based on wavelets, envelope curves, and time fronts. Figure 2 shows some of the family of minimum time routes on the time surface integrated about New York City for DC-8 jet aircraft flying to Europe at a constant pressure altitude of 300 millibars on October 17, 1960, and figure 2A shows how to create such a surface and attendant paths.

When portrayed in full, the situation depicted in figure 2 reveals cusps or focal points in the least time routes and thus provides alternate equal minimum-time paths to certain destinations. Such a situation refers to the induction of caustics and occurs at the rear of a closed vortex associated with a strong high or low or in a region of strong wind shear as in a jet stream or a frontal zone.

Under certain pressure distributions complicated systems of time fronts and routes exist, and, as with their isomorphic counterparts in optics, reflection, diffraction, refraction, and other patterns can be ascertained to exist in addition to, and in agreement with, the caustics.

With departure point W in figure 2A as the center, draw a circle with a radius representing the distance the aircraft can fly in one hour at its cruising true air speed. Only an arc of this circle is shown here, covering the general direction of the intended flight. This arc of the circle, labeled A1, indicated the maximum distances the aircraft could fly in the complete absence of wind. The effect of wind is estimated by drawing the appropriate wind vectors representing velocity and direction from a

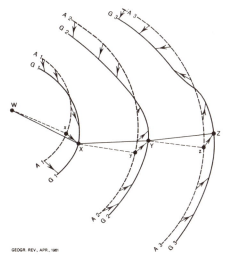

Figure 2A. Graphical determination of least time path.

number of points on the air position line A1. The heads of these wind vectors can be connected by the smooth curve G1, which designates the position of the time front after one hour and gives the farthest ground positions the aircraft could achieve in that time in the existing wind field.

From as many points as desired on curve G1, arcs may be swung off each with a radius again equal to the true air speed. The curve A2 is drawn as the envelope of these arcs. Displacement of A2 by the appropriate wind vectors yields G2, the position line of the time front at the end of the second hour after departure.

This procedure can be continued until the time front closest to required destination Z is obtained. In the case portrayed here destination lies exactly on a time front. If, as is usually the case, destination does not lie on a time front, interpolation will yield an adequate estimate of the least time in which the flights can be accomplished.

To obtain directly the specific route from W to Z that affords this minimum flight time, plot backward from destination to departure point. From Z (in this case on G3) the wind vector is plotted in reverse, giving the point z on A3. From this point the perpendicular is dropped to G2. This new point, Y, is on the required optimum flight path.

Repetition of this procedure yields point X on G1 and finally W as the departure point. The optimum flight path, here the minimum-time path, is then the curve WXYZ. This course represents the connected series of ground-speed-true-course vectors each perpendicular to its attendant air position curve. Likewise, the true-air-speed-true-heading vectors are perpendicular to their ground position curves, the time fronts.

The methods used here result in approximations of the desired result. Such graphical procedures involve arbitrary discontinuities and averages. Specifically, the accuracy achieved apart from the adequacy of the meteorological forecast tends to vary inversely with the length of the time interval chosen to portray wind velocities, air speeds, ground speeds, and the successive positions of the time front.

The final examples utilize a potential of population map based on 1960 census counts for the conterminous United States and computed and compiled jointly by the American Geographical Society and Princeton University using the IBM 7090 computer. Figure 3 shows this map, the most detailed such map yet produced. Potential of population is shown as a macrogeographic spatially continuous surface by means of isolines indicating geographical variation in the values on this surface. Potential of population measures the aggregate accessibility of the entire population of the country to all points when the value at any given point represents the summed contributions from all members of the population when each contribution is proportional to the reciprocal of the distance of the person away from the given point. More specifically, the value of potential of population at a given point is $\int \frac{D}{r} \, \mathrm{d}A$ when D is the population density

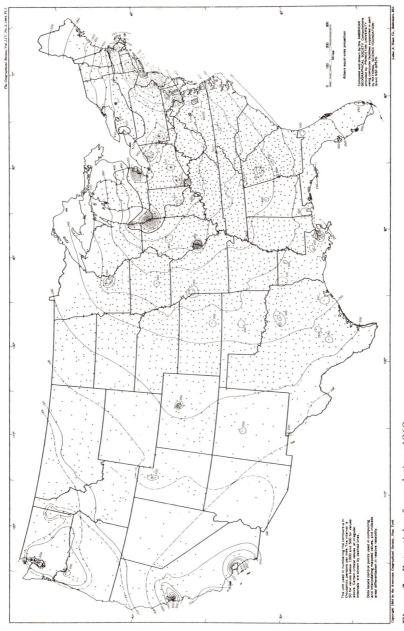

Figure 3. Potentials of population, 1960.

over any infinitesimal element of area, dA, and r is the distance of each such element from the given point. The integration is extended to all elements when D is not zero. The value of potential of population can be determined for as many points as required to facilitate mapping the surface by the contouring technique. Units of potential of population are in persons times distance to the minus one power. On figure 3 values are in thousands, persons per mile. To convert to thousands, persons per kilometer, the mapped values may be multiplied by the factor $0 \cdot 622$.

Maps of this sort indicate spatial structuring of a wide range of economic and social phenomena and are based upon formalization arising from empirical formulas describing certain spatial processes, i.e. flows that are of economic and social importance.

With regard to spatial structure it is important to note that values of non-urban land in the United States vary directly with and in close agreement with the potential of population surface. Specifically for 1960, land value in dollars per acre $= 6 \cdot 104^{-6}$ times potential of population raised to the $1 \cdot 6$ power when potential was in units of thousands, persons per mile. The coefficient of correlation was $0 \cdot 863$ based on state averages and thus was quite high. For the number of degrees of freedom obtaining here any value of the coefficient of correlation exceeding $0 \cdot 288$ is to be deemed significant at the five per cent fiducial level.

Urban land values increase with potential to an even higher power than for non-urban land so that one simple linear logarithmic function is inadequate. Even so, however, one may closely approximate a continuous land value surface for the United States by transforming the potential surface everywhere by the factor of proportionality and exponent of potential given above due to the peaking of potential locally in urban areas. Such a procedure will permit us to provide the illustration intended below.

Let us imagine that it is the task to establish from any given place the routes to all other places in the country and in each case the route is the one for which the total land acquisition cost would be at a minimum. A map of such a family of routes and the attendant isocost surface on which these routes represent gradient paths can be produced by means of the simple Huygens' graphical method. Figure 4 represents an isocost surface integrated about Lewistown, Montana. Lewistown, in this case, is a "sink" and all other places are relative "sources." The minimum land acquisition cost route to any other given point in the United States may be found by plotting the orthogonal trajectory to the isocost lines from the given point back to Lewistown. Hence, the routes are gradient paths. Several such paths are shown on the map. Note the divide between the northerly and southerly routes emanating from this town in Montana. To further emphasize the nature of such surfaces and the paths on them, an additional graphical presentation is supplied in figure 5 based on Murfreesboro, Tennessee.

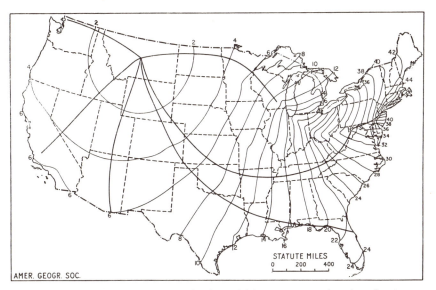

Figure 4. Cost surface and minimum land acquisition cost routes based on Lewistown, Montana (in hundreds of thousands of dollars).

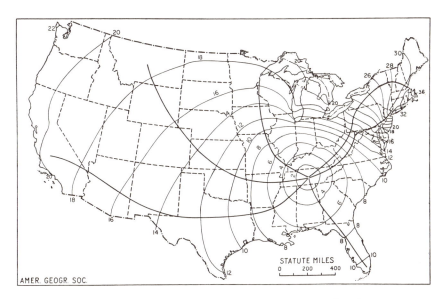

Figure 5. Cost surface and minimum land acquisition cost routes based on Murfreesboro, Tennessee (in hundreds of thousands of dollars).

Note particularly by comparing figures 4 and 5 that the path from Lewistown to Murfreesboro is precisely the same as that from Murfreesboro to Lewistown. Thus, although the sets of iso-cost contours differ greatly the tangents to both sets of contours coincide along this path. Because the original values on the land value surface are independent of the directions of lines through points on that surface, the conditions obtains that between any two points the least cost paths coincide. Other than this, however, a separate map is required for each point to obtain the minimum cost paths from all other points to the given point. The possibility of producing one general map to accomplish all such paths is being investigated.

For the minimum time paths for aircraft as shown in figure 2 the same path does not exist in opposite directions between any two points. As noted earlier, the effect at a point of the wind on the speed and the heading of an aircraft depends upon the heading and the speed with which the aircraft approaches the point.

The procedure used to achieve the surfaces represented in figures 4 and 5 involved a decision as to the finite width assumed for the minimum cost routes to be shown. This is indicated in the legend of figure 6 showing the graphical method employed to determine these surfaces and paths.

Many other examples might be given in addition to the three kinds given above, but the distance surfaces, time surfaces, and cost surfaces that these represent are indicative of the wide application of the basic ideas of

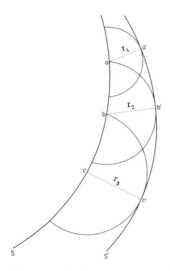

Figure 6. Graphical method for determining cost surface about a point. S and S' represent successive isocost lines; r_1, r_2, r_3 = distance radii achievable for outlay of $100000 assuming that 15 acres of land are required for each mile of highway length.

surfaces and paths in geography generally. This, of course, includes a part of the general theory of geodesics, and intensive study and the *literal* application of it to the real earth and especially the various conceptual surfaces covering it seems called for at this stage in the development of geography and its growth toward a truly unified discipline.

We have noted that mathematical and graphical solutions exist for our kinds of problems. In some cases, the mathematics is intractable, however, and graphical solutions, or rather approximations, do not exist or have not yet been invented.

A third kind of solution exists as an extension of the graphical method and that is to build three dimensional models of the various surfaces. For example, one could make a model of the equatorial Mercator projection of the iso-distance surface based on London. A convenient scale could be established between values on the distance surface and height above the plane on which London was regarded to exist. Thus London would become the pit with all others points elevated above it and with heights increasing in a monotonic and continuous fashion outwards until the one peak on the surface, the antipodal point for London was reached.

One could make this into an operational model, indeed a geographical analogue computer, for finding the great circle path on the surface from any point through London by allowing a ball to roll freely on the surface from that point. It will roll "down hill" to the pit of London along the gradient path. A difficulty exists because the ball has mass and gained momentum may cause it to depart from the locally steepest slope at some places. However, if the model's vertical scale is selected so that no slopes are too steep the problem is minimized. Introducing additional friction to the surface could help also.

Boundary problems exist also. The interruption of the map along a meridian is easily overcome by making the model periodic in an east-west direction. The fact that the geographical poles lie at infinity poses another less easily solved problem.

Models for the other phenomena presented in this paper are possible too. Thus, elevation on the surface could be made proportional to time from New York, or cost from Lewistown, etc. Our geodesic-finding ball would approximate the required paths.

Geography, geometry, and graphics which had their first grand synthesis in cartography at the time of Ptolemy stand to benefit mutually in the recently established cooperation which greatly extends and intensifies those early benefits. This time spatial patterns in general are being considered with regard to social, economic, and cultural phenomena as well as physical phenomena with recognition that the *geo* in geometry and the *geo* in geography have more in common than we would have dared to dream a decade ago.

References and acknowledgements

Part of this paper was presented to the International Geographical
Congress, London, July 1964. The topics in it were also presented to the
faculty and students of the Michigan Inter-University Community of
Mathematical Geographers in November 1964. The author is deeply
indebted to that group for their incisive comments and recommendations.

All figures have been drawn by the cartographic staff of the American
Geographical Society. Figures 1, 1A, 4, and 5 appear here for the first
time. The others have appeared previously in the *Geographical Review*
and are reproduced here with permission.

Thanks are owed to the Editor of the *Geographical Review* for allowing
the author to quote passages from his article, "Transatlantic Flights and
Pressure Patterns", from Vol. 51, No. 2, 1961, pp. 187–212 of that
journal.

John Z. Stewart in "The Use and Abuse of Map Projections,"
Geographical Review, Vol. 33, No. 4, 1943, pp.589–604, pointed out the
analogy between the inverse scale on any conformal map and the index of
refraction in geometrical optics.

Various staff members at the American Geographical Society have
advised the author. In particular, O. M. Miller, the assistant director of
the Society, offered valuable advice.

At Princeton University Professor Steve Slaby and Mr. C. Ernesto
Lindgren of the School of Engineering's Department of Graphics and
Engineering Drawing have provided many enlightening comments. They
are much interested in the modern revival of the cooperative assistance of
geometry and geography and especially of the role that graphics plays in
this.

Also at Princeton University, the sculptor in residence, Joe Brown,
advised the author and demonstrated for him certain of the techniques of
three-dimensional model building. It was in Mr. Brown's studio and with
his advice that the author completed the three-dimensional model of the
1960 Potential of Population Surface for the United States. This model
is currently included in the American Geographical Society's exhibit at
the New York World's Fair.

The gentlemen cited above must in no way be held accountable for the
defects of this paper, however. Those came from the mind of the author
alone. But, the author is deeply indebted to these many gentlemen and
would like to make known his gratitude to them.

Reprinted from *Geographical Review* **53**, 59–78, 1963

GEOGRAPHIC AREA AND MAP PROJECTIONS*

WALDO R. TOBLER

A BASIC truism of geography is that the incidence of phenomena differs from place to place on the surface of the earth. Theoretical treatises that assume a uniformly fertile plain or an even distribution of population are to this extent deficient. As Edgar Kant[1] has put it: "The theoretical conceptions, based on hypotheses of homogeneous distribution must be adapted to geographical reality. This implies, in practice, the introduction of corrections with regard to the existence of *blank districts, deserts of a phenomenon, massives* or *special points*. That is to say that in practice we have to take into especial consideration the anisotropical qualities of the *area geographica*." The *ceteris paribus* assumptions that are repugnant to a geographer are those which conflict seriously with the fundamental fact that the distribution of phenomena on the surface of the earth is highly variable. Von Thünen,[2] for example, postulates a uniform distribution of agricultural productivity; his economic postulates are no less arbitrary, but they disturb the geographer somewhat less. Christaller's central-place theory is in a similar category; for the necessary simplifying assumptions, among them a uniform distribution of purchasing power, are unsatisfactory from a geographic point of view.[3] In order to test the theory empirically, one must find rather large regions in which the assumptions obtain to a fairly close approximation. The theory can, of course, be made more realistic by relaxing the assumptions, but this generally entails an increase in complexity. An alternate approach, hopefully simpler but equivalent, is to remove the differences in geographic distribution by a modification of the geometry or of the geographic background. This has been attempted by other geographers with some success, but without clear statement of the problem.

* The author wishes to express his appreciation to the members of the Department of Geography, University of Washington, for their comments. Special stimulus was provided by Drs. John C. Sherman, William L. Garrison (Northwestern University), and William W. Bunge (Wayne State University). The National Science Foundation, through its program of Graduate Fellowships, provided financial support for a part of this study. The conclusions, of course, remain the author's responsibility.

[1] Edgar Kant: Umland Studies and Sector Analysis, *in* Studies in Rural-Urban Interaction, *Lund Studies in Geography*, Ser. B, Human Geography, No. 3, 1951, pp. 3–13; reference on p. 5. Italics are Kant's.

[2] Johann H. von Thünen: Der isolierte Staat in Beziehung auf Landwirtschaft und National-Ökonomie (Hamburg, 1826).

[3] Carlisle W. Baskin: A Critique and Translation of Walter Christaller's *Die zentralen Orte in Süddeutschland* (unpublished Ph.D. dissertation, Department of Economics, University of Virginia, 1957).

➤ Dr. Tobler is assistant professor, Department of Geography, University of Michigan, Ann Arbor.

60   THE GEOGRAPHICAL REVIEW

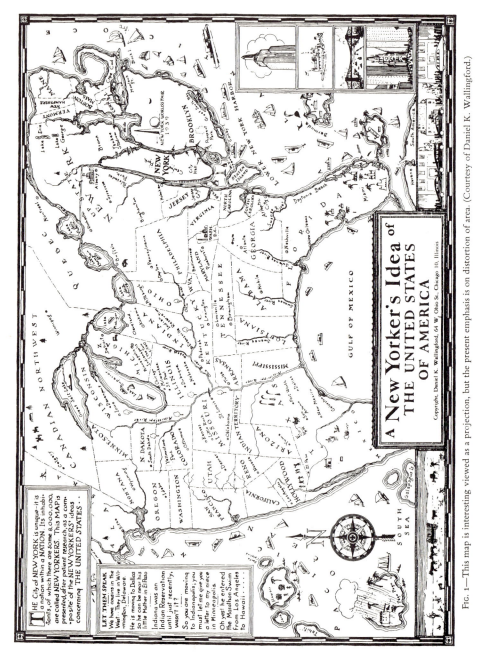

FIG. 1—This map is interesting viewed as a projection, but the present emphasis is on distortion of area. (Courtesy of Daniel K. Wallingford.)

Map projections always modify certain geometric relations and hence would seem well suited to the present task. However, instead of considering the earth to be an isotropic closed surface (as is traditional in the study of map projections), account can be taken of an uneven distribution of a phenomenon on this surface—the *area geographica*. The topic is approached by an

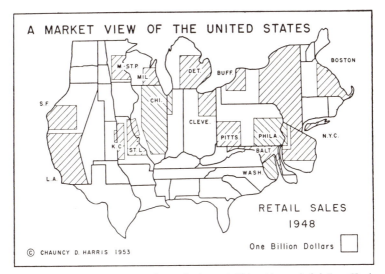

FIG. 2—Sizes of states proportional to retail sales, 1948. Major cities are shaded. From Harris, The Market as a Factor in the Localization of Industry [see text footnote 5 below], p. 320. (Courtesy of Chauncy D. Harris.)

examination of a number of published maps called cartograms in current cartographic parlance. Attention is here directed toward those types of cartograms which appear amenable to the metrical concepts of the theory of map projections, with no attempt at definition of the rather vague term "cartogram."

EXAMPLES OF CARTOGRAMS

Cartograms are of many types. The accompanying illustration showing "A New Yorker's Idea of the United States of America" (Fig. 1) contains several interesting notions. The thesis of cognitive behaviorism suggests that people behave in accordance with the external environment, not as it actually is, but as they believe it to be.[4] In this vein, the cartogram presented can be

[4] Harold and Margaret Sprout: Environmental Factors in the Study of International Politics, *Journ. of Conflict Resolution*, Vol. 1, 1957, pp. 309–328.

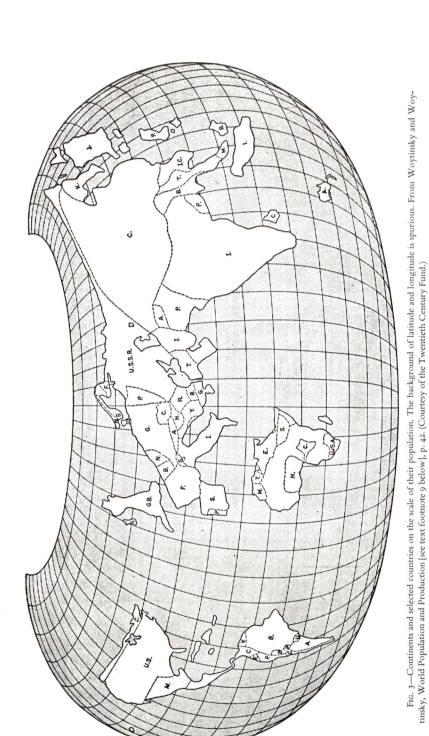

Fig. 3—Continents and selected countries on the scale of their population. The background of latitude and longitude is spurious. From Woytinsky and Woytinsky, World Population and Production [see text footnote 9 below], p. 42. (Courtesy of the Twentieth Century Fund.)

considered to illustrate one type of psychological distortion of the geographic environment that may occur in the minds of many persons. It is clear that the distortion is related to distance. Furthermore, the areas of the states are not in correct proportion; Florida, for example, appears inordinately large. Hence distortion of area can be recognized, though a complete separation of the concepts of distance and area is not possible in this instance.

The second illustration is also a distorted view of the United States (Fig. 2), but the purpose of this cartogram is somewhat different. The areas of the states and cities are shown in proportion to their retail sales, rather than in proportion to the spherical surface areas enclosed by their boundaries. Harris' point is that the expendable income, not the number of square miles, is a more proper measure of the importance of an area—at least for the purposes of the location of economic activity. Harris also presents cartograms of the United States with map areas of the states in proportion to the number of tractors on farms and to the number of persons engaged in manufacturing.[5] Raisz[6] presents a cartogram with the areas of the states in proportion to their populations. Hoover[7] stresses a point of view similar to that of Harris and presents a different cartogram of the United States, with map areas of the cities and states in proportion to their populations. Weigert[8] recognizes that the importance of a country may be more directly proportional to its population than to its surface area and presents a cartogram placing the countries of the world in this perspective. Woytinsky and Woytinsky[9] make extensive use of a similar cartogram, reproduced here as Figure 3. Zimmermann[10] presents further examples—cartograms of world population and of output of steel by country.

Whether all these cartograms are to be considered maps, based on projections, is a matter of definition and, as such, is not really important. Raisz

[5] Chauncy D. Harris: The Market as a Factor in the Localization of Industry in the United States, *Annals Assn. of Amer. Geogrs.*, Vol. 44, 1954, pp. 315–348. See also Chauncy D. Harris and George B. McDowell: Distorted Maps, A Teaching Device, *Journ. of Geogr.*, Vol. 54, 1955, pp. 286–289.

[6] Erwin Raisz: The Rectangular Statistical Cartogram, *Geogr. Rev.*, Vol. 24, 1934, pp. 292–296, Fig. 2 (p. 293).

[7] Edgar M. Hoover: The Location of Economic Activity (New York, Toronto, London, 1948), Fig. 5.6 (p. 88).

[8] Hans W. Weigert and others: Principles of Political Geography (New York, 1957), Fig. 9-2 (p. 296).

[9] W. S. Woytinsky and E. S. Woytinsky: World Population and Production (New York, 1953), pp. lxix–lxxii and 42–43, and *passim*.

[10] Erich W. Zimmermann: World Resources and Industries (rev. edit.; New York, 1951), p. 97. Another cartogram can be seen in David Greenhood: Down to Earth: Mapping for Everybody ([rev. edit.] New York, 1951), p. 236. The Library of Congress map collection also contains a large number of maps of this type.

stresses the point that his rectangular statistical cartograms are not map projections. The network of latitude and longitude on the Woytinskys' population cartogram (Fig. 3) suggests a map projection but is actually spurious, as the Woytinskys themselves remark. However, since all maps contain distortion, the diagrams can be regarded as maps based on some unknown projection. Certainly the definition that considers a map projection to be an orderly arrangement of terrestrial positions on a plane sheet suffices. It also seems adequate to demonstrate that diagrams similar to these cartograms can be obtained as map projections. But what is the nature of these projections? No such map projections are given in the literature of the subject. The question is approached by a detailed examination of a simpler problem posed by Hägerstrand.

Hägerstrand has been concerned with the study of migration. In discussing the cartographic problem, he states:[11]

> The mapping of migration for so long a period, giving the exchange of one single commune with the whole country in *countable* detail, cannot be made by ordinary methods. All parts of the country have through the flight of time been influenced by migration. However, different areas have been of very different importance. With the parishes bordering the migrational centre, the exchange has numbered hundreds of individuals a decade. At long distances only a few migrants or small groups are recorded. A map partly allowing a single symbol to be visible at its margin, partly giving space to the many symbols near its centre, calls for a large scale since *we wish to be able to count on the map.*

It is desired to count symbols on the map. This is a clear statement of a common cartographic problem. The situation occurs frequently in the mapping of population, where high concentrations appear in restricted areas and smaller numbers are spread more thinly throughout the remainder of the map. Certainly every cartographer has at some time wished for a distribution of a phenomenon which did not seem to require that all the symbols overlap. One solution has been the introduction of so-called three-dimensional symbols.[12] An alternate solution is here suggested, based on the theory of map projections. Also note the distinction between the common geographic use of an equal-area map to illustrate the distribution of some phenomenon and Hägerstrand's emphasis on the recovery of information recorded on the map.

In the problem as formulated by Hägerstrand, the exchange of migrants is known not to be distributed arbitrarily but is a function of distance from a

---

[11] Torsten Hägerstrand: Migration and Area, *in* Migration in Sweden, *Lund Studies in Geography*, Ser. B, Human Geography, No. 13, 1957, pp. 27–158; reference on p. 73. Italics are Hägerstrand's.

[12] Cf. Arthur H. Robinson: Elements of Cartography (2nd edit.; New York and London, 1960), Figs. 9.16 (p. 169) and 9.17 (p. 170).

center, the commune being studied. More commonly, differences from one area to another vary much more irregularly, as in, for example, the distribution of population throughout the world. Careful reading of Hägerstrand's statement suggests that the functional dependence is one of decreasing migratory exchange with increasing distance from the center. This can be rec-

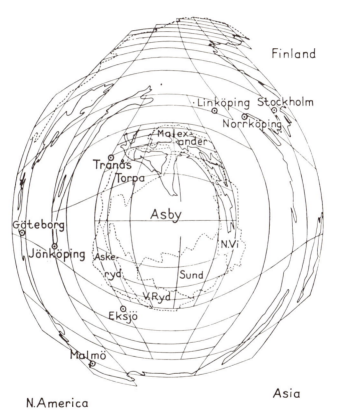

FIG. 4—Hägerstrand's logarithmic map. The numbers and grid lines refer to the Swedish plane coordinate system. From Hägerstrand, Migration and Area [see text footnote 11 above], p. 73 (Courtesy of Torsten Hägerstrand.)

ognized as a simple distance model often employed by geographers. In particular, the suspected function of distance can be postulated to be continuous and differentiable, strictly monotone decreasing, and independent of direction. If these postulates are accepted, the functional dependence can be shown on a graph as a continuous curve, in this instance a curve of negative slope. The curve can be considered a profile along an azimuth, and the expected incidence of migration could be shown on a map by isolines. This suggests that variants of the solution to Hägerstrand's problem can be applied to many isoline maps. Population density, for example, is often illustrated by isolines drawn on maps, and an approach to the population cartograms is suggested. Hägerstrand's own solution is as follows:[13]

> The problem is solved by the aid of a map-projection in which the distance from the centre shrinks proportionally to the logarithm of the real distance. (The method was suggested to the author by Prof. Edg. Kant. Maps of a similar kind are used for the treatise "Paris et l'agglomération Parisienne" 1952.) The rule obviously cannot be applied to the shortest distances. Thus the area within a circle of one km radius has been reduced to a dot.
> The distortion in relation to the conventional map is of course considerable.

The basis for the choice of the logarithmic projection (Fig. 4) is not clearly indicated in this statement. An azimuthal projection that yields the desired result seems to have been plucked out of thin air. Working backward, however, the radial scale distortion is seen to be ρ^{-1} (where ρ is the spherical distance), and it can be inferred that the projection was obtained by taking the suspected function of distance as the radial scale distortion, as can be done for any of the distance models employed by geographers.[14] The space elimination at the origin is appropriate, for it excludes the commune being studied (which does not belong to the domain of migration). But is Hägerstrand's the most valid solution to the problem as formulated? The concept of primary concern is not distance but area. This is implicit in the statement that it is desired to be able to count symbols on the map. The suggestion is that the map show the areas near the center at large scale and those at the periphery at small scale. Such maps would be useful in most studies of nodal regions. Hägerstrand's solution achieves this objective, as can be verified by calculation of the areal distortion, at least for areas near the center of the map. But so do the orthographic projection, the square-root projection, and many others. The azimuthal equidistant centered on the antipodal point also yields the desired solution and has been used for this purpose by Michels.[15] Kagami[16] suggests an alternate solution when faced with an almost identical problem.

---

[13] Hägerstrand, *op. cit.* [see footnote 11 above], p. 74. The reference is to P.-H. Chombart de Lauwe and others: Paris et l'agglomération parisienne (2 vols.; Paris, 1952).

[14] For further details on this procedure see W. R. Tobler: Map Transformations of Geographic Space (unpublished Ph.D. dissertation, University of Washington, 1961), pp. 114–117.

[15] F. W. Michels: Drie nieuwe kaartvormen, *Tijdschr. Kon. Nederl. Aardrijksk. Genootschap*, Ser. 2, Vol. 76, 1959, pp. 203–209. See also D. M. Desoutter: Projection by Introspection, *Aeronautics*, Vol. 40, 1959, pp. 42–44.

[16] Kanji Kagami: The Distribution Map by the Method of Aeroview, *Geogr. Rev. of Japan*, Vol. 26, 1953, pp. 463–468 (with English abstract).

Charts for aircraft pilots have also been prepared using maps that have a large scale near the center and a small scale at the periphery.[17]

CARTOGRAMS AS PROJECTIONS

To clarify the situation, one should note that it is the areal scale, and not the linear scale, which is important. Furthermore, it is natural to require that the areal distortion be *exactly* the same as the expected or observed distribution. Somewhat more precisely, Hägerstrand's problem can be generalized in the following manner. In the domain under consideration there are locations from which migration to the center originates. If we consider the beginning point of each migration to be an "event," each small region (element of area) will (or is likely to) contain a certain number of events. Hence with each proper partition of the domain is associated a number, and the area contained within the boundaries of the corresponding partition on the map is to be proportional to this number. The similarity to the cartograms previously presented is now clearer. In each instance a set of non-negative numbers (people, dollars) has been associated with a set of bounded regions (cities, states, nations). The objective is to display the regions on a diagram in such a manner that the areas within the boundaries of the regions on the diagram are proportional to the number associated with the particular region. Harris recognizes the similarity of the concepts, for his cartogram "A Farm View of the United States"[18] is accompanied by a histogram of the number of tractors by states. On an equal-area projection, the number associated with each partition is the spherical (or ellipsoidal) surface area.

There seem to be two methods of attacking the details of these map projections. One assumes differentiability; the other is an analogue of the first but employs what might be called rule-of-thumb procedures. Each method has advantages and disadvantages. The differentiable cases display the similarity to equal-area map projections somewhat better, whereas the approximation methods are simpler to use with empirically obtained data. The differentiable cases also allow explicit solution for the pair of functions necessary to define a map projection. No attempt is made here to duplicate the specific cartograms illustrated; the purpose is only to indicate the class of projections to which they belong.

The data are somewhat difficult to manipulate when the partitions of area are large. It is therefore convenient to reduce the values associated with

---

[17] [Leslie Y. Dameron:] Terminal Area Charts for Jet Aircraft, *Military Engineer*, Vol. 52, 1960, p. 227.

[18] Harris, *op. cit.* [see footnote 5 above], p. 338.

68   THE GEOGRAPHICAL REVIEW

each portion of the domain to density form, and to think in terms of a continuous (integrable and differentiable) distribution that can be represented by isolines on a sphere. The details of this device are well known and can be omitted here.[19] The map area between given limits is then to be proportional to the total distribution between corresponding limits. The density distribution on the surface of a sphere is assumed to have been described by an equation. For equal-area projections the density of spherical surface area is always constant (unity), so that correct values are also obtained in this

FIG. 5—This illustration can be considered as either (a) isolines of population density or (b) polar coordinates after a transformation.

special situation. As is true of area, finite densities sum to a finite value, so that the density-preserving property of the projections to be achieved obtains both locally and in the large. The use of density values also facilitates the further objective that common boundaries between regions should again coincide on the final map.

The derivation of the cartograms under consideration as map projections follows directly from the preceding discussion. A mathematical analysis of this class of map projections is given in the Appendix. A special case, of some practical interest, is given here to illustrate the general method.

The distribution of population in an urban area can be described as a density function $D(\delta, \gamma)$ on a plane, using polar coordinates δ, γ. Horwood[20] has suggested one such distribution in which the density decreases monotonically from the center but also varies from one direction to the next (Fig. 5). The specific theoretical function taken by Horwood is such that the density is highest along symmetrically spaced radial streets (n in number) and less in the interstitial areas, which is not unrealistic and is easily described by trigonometric functions or Fourier Series. The population is then given by the integral

$$\iint_R \delta D(\delta, \gamma) d\delta d\gamma \ .$$

[19] See C. E. P. Brooks and N. Carruthers: Handbook of Statistical Methods in Meteorology, *M.O. 538*, London, 1953, pp. 161–165; or Calvin F. Schmid and Earle H. MacCannell: Basic Problems, Techniques, and Theory of Isopleth Mapping, *Journ. Amer. Statist. Assn.*, Vol. 50, 1955, pp. 220–239.

[20] E. M. Horwood: A Three-Dimensional Calculus Model of Urban Settlement (paper presented at the Regional Science Association Symposium, Stockholm, August, 1960).

AREA AND PROJECTIONS   69

To transform this to the map plane so that all map areas have identical densities, set

$$\iint_{R'} r\, dr\, d\theta \;=\; \iint_{R} \delta D(\delta,\gamma)\, d\delta\, d\gamma \;,$$

or

$$\iint_{R'} r|J|\, d\delta\, d\gamma \;=\; \iint_{R} \delta D(\delta,\gamma)\, d\delta\, d\gamma \;,$$

which is equivalent to

$$r|J| \;=\; \delta D(\delta,\gamma) \;,$$

where

$$\pm J \;=\; \frac{\partial r}{\partial \delta}\,\frac{\partial \theta}{\partial \gamma} \;-\; \frac{\partial \theta}{\partial \delta}\,\frac{\partial r}{\partial \gamma} \;.$$

For one solution, not necessarily the most appropriate but simple, stipulate that the transformation is to be azimuthal, that is, that $\theta = \gamma$. Then

$$\frac{\partial \theta}{\partial \delta} = 0 \;,\quad \frac{\partial \theta}{\partial \gamma} = 1 \;,\text{ and } J = \frac{\partial r}{\partial \delta} \;.$$

The equation to be solved for r is consequently

$$r^2 \;=\; 2\int \delta D(\delta,\gamma)\, d\delta \;+\; g(\gamma) \;,$$

and the remaining details are matters of integration and root extraction. This example could be extended to a sphere or spheroid, but for an urban area there is little point in such extension. The image of the original polar coordinates on the final map might appear as shown in Figure 5.

Although further details are in the Appendix, certain results from the mathematical analysis are worth noting here. It is easily shown that the transformations are a generalization of equal-area projections in the sense that *all* equal-area projections represent a special case. Moreover, this class of projections can be obtained by setting Tissot's measure of areal distortion equal to the given (expected, probable) density distribution. It is also apparent that there are an infinite number of solutions for any specific density. This suggests that additional conditions be applied. Of the many possible conditions, two are of particular interest. Since this class of projections is equivalent to projections with areal distortion, and since all conformal projections of a sphere distort area, it follows that a conformal projection with a specific areal distortion should yield a solution. The transformation also may be taken so that cost or time distances from the map center are correctly represented.

Occasionally the assumption of continuity of a distribution is not warranted. The data are often in the form of discrete locations, as on a population dot map; or are grouped into areal units, such as census tracts; or refer to areal units rather than to infinitesimal locations, such as land values that refer to specific parcels of land. In these cases an analytic solution usually is not feasible and rule-of-thumb approximations are useful. Even in the case of continuous distributions, descriptive equations are difficult to obtain and, at present, are not available for geographic data, though theoretically possible. Approximation methods, therefore, are useful. They can also be used to demonstrate some of the different types of particular solutions available and some of the additional conditions that may be applied. The approximation methods are no less valid than the methods used in the differentiable cases and can be formalized to the same extent, but they are more akin to topological transformations than to those traditionally associated with cartography.

The only known description of the method used in the preparation of the cartograms previously mentioned is that given by Raisz;[21] the method used by others is presumably similar. The populations of the states are taken as given, and rectangles proportional to population are drawn on a sheet of paper; adjacent rectangles are adjusted until neighbor relations and overall shape are approximately correct. This is illustrated in Figure 6. Though the example is very simple, there are still an infinite number of solutions, but some seem more appropriate than others. Preservation of the internal topology is one condition that seems desirable; this is in fact a requirement that the map (not the distribution) be continuous (a homeomorphism—neighborhoods are preserved under the mapping). Preservation of the shape of the external boundary is another condition that might be applied. Alternately, one might wish the boundary to map into a specific shape. These last two conditions are difficult to specify even in the analytic case.

If one thinks in terms of a map of a part of the earth's surface, an obvious difficulty is that the immediately foregoing examples do not indicate where positions within the original areal units are to be placed within the corresponding partitions of the transformed image. Stated in another way, if locations in the original are described by latitude and longitude, where are the images of these lines in the transformed image? If the partitions represent states, the placement of cities is rather arbitrary, and so on. Here the differ-

---

[21] Erwin Raisz: Rectangular Statistical Cartograms of the World, *Journ. of Geogr.*, Vol. 35, 1936, pp. 8–10; *idem*, The Rectangular Statistical Cartogram [see footnote 6 above].

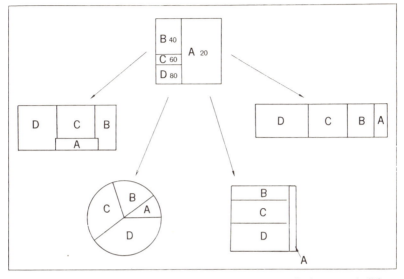

Fig. 6—Sample transformations of a unit square with different amounts of a phenomenon in different portions illustrating several of the possible solutions.

entiable cases display a distinct advantage. However, if a coordinate system is introduced in the original, and an assumption of uniform density within each partition (for example, states) is made, the difficulty can be circumvented by estimating lines of equal increments of density on the original. These lines then correspond to an equal-area grid system on a plane, and the converse. A similar method can be employed when the original data are given in the form of a dot map. If a partition has no entries, the map area should vanish, a collapsing of space or a many-to-one mapping. Figure 3 actually consists of several domains; otherwise, ocean areas would be eliminated (lines of latitude and/or longitude coincide), just as Greenland and Antarctica do not appear on the map. Although there is some population in the ocean areas, the amounts are so small as to be negligible. In the continuous case with zero density the transformation becomes many-to-one (a collapsing of space) for this part of the domain.

The approximation methods need not be discussed in more detail; they are fairly simple and do not reveal information that is not readily apparent from an examination of the equations given in the Appendix. More interesting, and more difficult to evaluate, are the geographic uses of maps obtained by the foregoing types of projections or transformations. These applications should also suggest the additional conditions to be applied in selection of a specific transformation from the infinite variety of particular solutions available.

GEOGRAPHIC APPLICATIONS

Obviously, the map projections obtained can be used as were the cartograms previously presented, for they were derived by consideration of such cartograms. These many applications need not be repeated. Further, any distribution plotted on a map using such a transformation shows a ratio; income symbolized on a map equalizing population density shows per capita income, and so on. The projections may likewise be useful as base maps in simulation or other studies in which data are plotted by computer.

It is also clear that any grid system which partitions the area of the plane map into units of equal size will yield a partitioning of the basic data into regions containing an equal number of elements when mapped back to the original domain. For example, states might be partitioned into electoral districts in such a manner that all districts contained an equal number of voters. The specific equal-area grids on a plane are infinite in number, so that this procedure is not really of much assistance. Equal-area grids are also difficult to define along irregular boundaries, and partitionings (electoral districts, and so on) are usually required to satisfy numerous additional conditions (coincide with city and county boundaries, and so on). To attempt to use the transformations in this manner seems politically impractical, though theoretically suggestive.

More interesting applications can perhaps be found in the theories of Von Thünen and Christaller. It is in this context that Harris and Hoover attempted to use their cartograms. Von Thünen assumes a uniform fertility of agricultural land, Christaller a uniform distribution of rural population or income, though both attempt to relax these unrealistic assumptions somewhat. If one postulates that agricultural fertility can be measured and varies from place to place—that is, that fertility can be described by a relation $F = f(\phi, \lambda)$—and if one then applies a transformation of the type described, areas of high fertility will appear enlarged. One can then plot[22] an even yield (for example, in bushels) per unit of map area and, using the inverse transformation, return to the original domain. The even distribution of yields will now be uneven, and in fact corresponds to the distribution of fertility. This becomes more interesting if one adds the condition that cost distances from (or to, but not

---

[22] The plotting can be conceptual, or it can be internal in a digital computer, and need not actually be performed.

both) a market place appear as map distances from the center of the map and that the intensity of use (yields) decreases with cost distance. That is, on the map transformed so that all areas appear of equal fertility, returns are to be plotted as decreasing from the center of the map, as in the Von Thünen model. The inverse transformation will then display a distribution of intensity of use that takes into account fertility and cost distance from the market place. The measurement of agricultural fertility is by no means easy. Dunn[23] doubts that such measurement can be achieved, but the United States Department of Agriculture publishes detailed information with a ranked classification (measurement on an ordinal scale) of rural land based on its economic value. Cost distances are used in the preparation of the map projection as another application of the notion that the earth should perhaps not be treated as an isotropic sphere. It is necessary to take into account not only the shape of the earth but also the realities of transportation on its surface. Automobiles, trains, airplanes, and other media of transport can be considered to have the effect of modifying distance relations—measured in temporal or monetary units—in a complicated manner. It can be shown (see Appendix) that a density-preserving projection with a continuous and monotonic but otherwise arbitrary centrally symmetric distance function can be obtained. This distance function can be the empirically obtained cost- or time-distance from the market place.[24]

Just as the Von Thünen model can be applied to cities,[25] the foregoing discussion can be rephrased using "suitability for construction" instead of fertility. Many urban areas are already built up, and construction is no longer feasible; other areas are blighted and have but little appeal; some locations have high prestige value; site and topographic factors vary; and so on. Undoubtedly, measurement of these values is difficult. Requirements for different classes of land use differ, and some measure of intensity of use seems required. Land costs are biased, since they reflect accessibility and an estimate of potential returns. Nevertheless, the transformation and its inverse can be used as before. Such a transformation takes into account only two factors and is therefore of only limited assistance in explaining the totality of urban land uses. The currently available models of urban structure are not outstandingly more successful.

---

[23] Edgar S. Dunn, Jr.: The Location of Agricultural Production (Gainesville, Fla., 1954), pp. 67–69.

[24] A more extensive discussion of this topic can be found in Tobler, op. cit. [see footnote 14 above], pp. 78–141.

[25] William Alonso: A Model of the Urban Land Market: Locations and Densities of Dwellings and Businesses (unpublished Ph.D. dissertation, University of Pennsylvania, 1960).

Christaller in his work on geographic location[26] assumes a uniform distribution of the underlying rural population and then obtains sets of nested hexagonal service areas and a hierarchy of cities regularly spaced throughout the
landscape. It has been shown how an uneven distribution may be made to
appear uniformly distributed, and the pertinent question is whether Christaller's resulting pattern will now be observed. The answer is difficult, for several reasons. Given an empirical distribution of income and market areas,
the transformation is to make the income densities uniform and to send the
market areas into hexagons. It is not clear how this latter condition is to be
specified in choosing a particular transformation from the infinite set. Christaller obtains hexagons from consideration of circular service areas, and it is
known that only the stereographic projection sends all circles into circles. The
stereographic projection, however, will certainly not result from the density-
preserving transformation in the general case. Conformal projections in
general preserve circles as circles, but only locally, and would require satisfying both conditions of conformality and a specific areal distortion. For
relatively small service areas conformal transformations may be suitable. The
solution (if one exists) to this problem is obscure. It is possible, of course, to
draw hexagons on a map of some region transformed in such a manner that
densities are uniform and, by use of the inverse transformation, to examine
the resulting pattern of curvilinear polygons in the original domain. There is
a slight problem here of specifying an initial orientation for the hexagons and
of fitting hexagons to the boundaries of the image region. The appearance of
the transformed hexagons will of course differ for each transformation in the
infinite set. Nevertheless, an experiment of this nature has recently been completed by Getis, using expendable income data for the city of Tacoma.[27]
Richardson's conformal transformations of hexagonal patterns are somewhat
similar.[28] Some such procedure is also implied by Isard's schematic diagrams
of a hypothetical landscape.[29] Conceptually, Isard's notions are correct, but the
boundaries of the service areas will almost certainly not be straight lines, as
they have been drawn in his illustrations. Conversely, one might use Vidale's
method of partitioning a landscape into service areas,[30] apply a transforma-

---

[26] Walter Christaller: Die zentralen Orte in Süddeutschland (Jena, 1933).

[27] Arthur Getis: A Theoretical and Empirical Inquiry into the Spatial Structure of Retail Activities
(unpublished Ph.D. dissertation, University of Washington, 1961), pp. 89–102.

[28] Lewis F. Richardson: The Problem of Contiguity (an appendix to his "The Statistics of Deadly
Quarrels" [Pittsburgh, Chicago, and London, 1960]), General Systems, Vol. 6, 1962, pp. 139–187.

[29] Walter Isard: Location and Space-Economy ([Cambridge and] New York, 1956), Figs. 52 (p. 272),
53 (p. 277), and 54 (p. 279).

[30] Marcello Vidale: A Graphical Solution of the Transportation Problem, Operations Research, Vol.
4, 1956, pp. 193–203.

tion, and examine the images of the service areas to see whether they resemble hexagons. Such an empirical experiment does not appear difficult; one can choose simple density distributions and use the simpler and more obvious transformations. None of these methods is as satisfactory as a theoretical solution, of course, though they may shed further light on the nature of the problem. Christaller's hexagons also need not be retained. Another approach is to consider threshold populations, not hexagons. From this point of view the boundaries of service areas overlap and are somewhat indeterminate. Adding the concept of the range of a good enables one to define the region in terms of cost distances. In this instance the useful map projections are those which make cost distances from some location proportional to map distances from that location and which distribute densities evenly.

Christaller is also concerned with distances; his circular service areas are more akin to geodesic circles using a "subjectively valued time-cost distance" (*sic*), and his spacing of cities stipulates some distance between cities. Yet distances are not preserved by the transformations; preservation of all distances is certainly not possible if densities are to be uniformly distributed on a plane map. Clearly, then, application of the suggested transformations to theories similar to those of Von Thünen and Christaller is difficult and only partly successful, though promising and capable of improvement. The deficiencies are to a certain extent due to the inadequacies of the theories themselves; for, at present, they are neither sufficiently general nor explicitly formulated.

Valuable map projections can be obtained that do not conform to the traditional geographic emphasis on the preservation of spherical surface area but rather distort area deliberately to "eliminate" the spatial variability of a terrestrial resource endowment. In many ways these maps are more realistic than the conventional maps used by geographers and would be of value even if the earth were a disk, as some ancients believed. The important point, of course, is not that the transformations distort area but that they distribute densities uniformly. It is hoped that future textbook presentations on the subject of map projections will include discussion of this interesting and highly useful class of transformations.

APPENDIX

1. The element of area on a locally Euclidean (but otherwise arbitrary) two-dimensional surface is given by the well-known formula due to Gauss:[31] $dA = (EG - F^2)^{1/2} \, du\, dv$. The

---

[31] See, for example, Dirk J. Struik: Lectures on Classical Differential Geometry (Reading, Mass., 1950), or any other text on differential geometry. Einstein's more convenient notation is not employed in cartography.

element of density on a surface is given by $dD = D(u,v)dA$, where $D(u,v)$ represents the given (expected, probable) value at the point u,v. The general problem hence reduces to one of finding u' and v' as functions of u and v to satisfy

$$\iint_{R'} (E'G' - F'^2)^{1/2} |J| du'dv' = \iint_{R} D(u,v)(EG - F^2)^{1/2} dudv \qquad (1.1)$$

or

$$(E'G' - F'^2)^{1/2} |J| = D(u,v)(EG - F^2)^{1/2} . \qquad (1.2)$$

For a sphere $EG - F^2$ is equal to $R^4\cos^2\phi$, using geographical coordinates ϕ and λ, or to $R^4\sin^2\rho$, using spherical coordinates ρ and λ. In the present instance the interest is only in plane maps; for a plane, $E'G' - F'^2$ is equal to 1, using rectangular coordinates x and y, or to r^2, using polar coordinates r and θ. The interesting cases will generally be oblique projections, but this requires only a relabeling.

When the Jacobian determinant (J) is written out in full, the following partial differential equations obtain:

$$\frac{\partial x}{\partial \lambda} \frac{\partial y}{\partial \phi} - \frac{\partial x}{\partial \phi} \frac{\partial y}{\partial \lambda} = \pm D(\phi,\lambda)R^2 \cos \phi , \qquad (1.3)$$

$$r\left[\frac{\partial r}{\partial \rho} \frac{\partial \theta}{\partial \lambda} - \frac{\partial r}{\partial \lambda} \frac{\partial \theta}{\partial \rho} \right] = \pm D(\rho,\lambda)R^2 \sin \rho . \qquad (1.4)$$

2. The difficulty of an explicit solution to 1.3 or 1.4 will depend on the specific form of the density function and the additional conditions applied. As is typical of differential equations, in general there will be an infinitude of particular solutions. Certain simple solutions, however, are immediately apparent. For example, if $\partial x/\partial \phi = 0$ and $\partial x/\partial \lambda$ is arbitrary, then

$$y = R^2 \int \frac{\pm D(\phi,\lambda) \cos \phi}{\partial x/\partial \lambda} d\phi + g(\lambda) . \qquad (2.1)$$

Or if $\partial y/\partial \lambda = 0$, and $y = f(\phi)$ is given, then

$$x = R^2 \int \frac{\pm D(\phi,\lambda) \cos \phi}{\partial y/\partial \phi} d\lambda + g(\phi) . \qquad (2.2)$$

In polar coordinates a similar procedure is available. Taking $\partial \theta/\partial \rho = 0$ and a given $\partial \theta/\partial \lambda$ yields

$$r^2 = 2R^2 \int \frac{\pm D(\rho,\lambda) \sin \rho}{\partial \theta/\partial \lambda} d\rho + g(\lambda) . \qquad (2.3)$$

An azimuthal projection is obtained if $\theta = \lambda$, conic projections if $\theta = n\lambda$, a truncated conic projection if $\partial \theta/\partial \lambda = n$, etc. Taking $\partial r/\partial \lambda = 0$, and with $r = f(\rho)$ selected arbitrarily, yields

$$\theta = R^2 \int \frac{\pm D(\rho,\lambda) \sin \rho}{r(\partial r/\partial \rho)} d\lambda + g(\rho) . \qquad (2.4)$$

3. The condition that a map of the sphere be equal-area can be written as

$$\frac{|J|}{R^2 \cos \phi} = 1 \text{ (or constant)} . \tag{3.1}$$

Hence it follows immediately that equal-area projections represent the special case $D = 1$ (or constant).

4. Areal distortion (S) is, by definition, the ratio of the element of area on the map to the element of area on the original. In other words,

$$S = \frac{dA'}{dA} = \frac{(E'G' - F'^2)^{1/2}}{(EG - F^2)^{1/2}} . \tag{4.1}$$

From a simple substitution it is seen that the density is the same as the areal distortion (i.e. $D = S$). In Tissot's notation $S = ab$, the product of the linear distortion in two orthogonal directions. Knowing this relation, we can obtain the desired transformations by choosing the areal distortion to match exactly the expected or known density distribution.

5. If the density is given by $\cos^{-4}(\rho/2)$ and an azimuthal projection is desired, equation 2.3 yields the stereographic projection. Although such a density is unlikely, this demonstrates the existence of conformal projections within this class of projections. The suggestion is that a conformal version exists among the solutions for many, if not all, non-constant densities. Though the areal distortion on conformal projections is easily calculated, the existence of conformal projections with a given areal distortion involves more subtle considerations, which are not presented here.[32]

6. According to Tissot, every non-conformal transformation retains as orthogonal curves one, and only one, pair of curves orthogonal on the original. An interesting question is whether the transformation can be determined so that the lines of latitude and longitude are the lines that remain orthogonal. For densities that depend on only one parameter the condition is readily obtained. For example, if $D = D(\phi)$ and $\partial x/\partial \lambda = 1$, equation 2.1 yields a cylindrical projection. Korkine's analysis of equal-area projections may be of use in obtaining the general case.[33]

7. Transport costs are often said to increase at a decreasing rate with distance, i.e. $\partial^2 r/\partial \rho^2 < 0$. If $r = f(\rho)$ and a density $D(\rho, \lambda)$ is given, equation 2.4 yields a solution that renders map distances proportional to transport costs and distributes densities evenly (see 8.4). An even more interesting result would be the simultaneous solution of 1.4 with an arbitrary $r = f(\rho, \lambda)$.

8. A few particular solutions may be of interest. From 2.3 an azimuthal projection for a linearly decreasing density $D = a\rho + b$, $a < 0 < b$, yields

$$r = [2R^2(- a\rho \cos \rho - b \cos \rho + a \sin \rho)]^{1/2} . \tag{8.1}$$

If the density distribution in Hägerstrand's problem is assumed to be ρ^{-1}, the appropriate azimuthal projection is

---

[32] See, for example, Richardson, *op. cit.* [see footnote 28 above], equation 4/54 (p. 158).

[33] A. Korkine: Sur les cartes géographiques, *Mathematische Annalen*, Vol. 35, 1890, pp. 588–604. Also note the similarity to equations derived by the Russian Urmaev, as discussed in D. H. Maling: A Review of Some Russian Map Projections, *Empire Survey Rev.*, Vol. 15, 1959–1960, pp. 203–215, 255–266, and 294–303; reference on pp. 210–213.

$$r^2 = 2R^2 \int \frac{\sin \rho}{\rho} d\rho = 2R^2 \left(\rho - \frac{\rho^3}{3 \cdot 3!} + \frac{\rho^5}{5 \cdot 5!} - \frac{\rho^7}{7 \cdot 7!} + \frac{\rho^9}{9 \cdot 9!} \cdots \right). \quad (8.2)$$

Additional azimuthal projections for densities equaling $\exp(-\rho)$ or $(2\pi)^{-1/2} \exp(-\rho^2/2)$ would appear to be of geographic interest, and are relatively easily obtained.

From 2.4 one obtains an equidistant version with $r = R\rho$ and $D = \pi - \rho$:

$$\theta = (-1 + \pi/\rho)\lambda \sin \rho . \quad (8.3)$$

Also from 2.4 but with $r = R(\rho)^{1/2}$, $D = \pi - \rho$, one has

$$\theta = 2\lambda (\pi - \rho) \sin \rho + g(\rho) . \quad (8.4)$$

In all these instances it is necessary to examine the resulting transformation for one-to-oneness. Choice of the constants of integration may be of importance. In some instances the substitution of difference equations for the differential equations may be appropriate. The author has calculated further special cases, which will be made available to interested parties.

9. It is suggested that these projections be referred to by their mathematical name; that is, as transformations of surface integrals.

Reprinted from *Strassenbau und Strassenverkehrstechnik*, **86**, 184–190, 1969

Wardrop

MINIMUM-COST PATHS IN URBAN AREAS

J. G. Wardrop

Research Group in Traffic Studies, University College London
London, United Kingdom

ABSTRACT

Previous work has shown that there is a direct analogy between routes which minimise transport cost and the paths of light rays. Following this analogy, the following equation is derived for a minimum-cost route:

$$\varkappa = \frac{1}{\mu}\frac{d\mu}{dn}\sin\varphi$$

where $\varkappa$ is the curvature, μ is the cost per unit distance, $\frac{d}{dn}$ denotes differentiation along the normal to the curve $\mu =$ constant, and φ is the angle between the path and the normal. If μ is increasing, the path curves towards the normal; if μ is decreasing the path curves away from it.

Conformal transformations can be used in a variety of cases to simplify the calculation of cheapest routes and curves of equal cost from a given origin. For cities with circular symmetry, transformations of the form $\omega = z^p$ can be used, corresponding to $\mu = cr^{p-1}$, where r = distance from city centre. The special case $\omega = \log z$ corresponds to $\mu = c \cdot r^{-1}$. Examples are worked out in terms of minimum journey times for the cases where speed is proportional to $r^{1/2}$ and to r.

RESUME

Itinéraire de coût minimum dans les villes

Une étude précédente a montré qu'il existe une analogie directe entre les itinéraires rendant minimum le coût du transport et les trajectoires des rayons lumineux. Suivant cette analogie on obtient l'équation suivante pour un itinéraire de coût minimum:

$$\varkappa = \frac{1}{\mu}\frac{d\mu}{dn}\sin\varphi$$

où $\varkappa$ est la courbure, μ est le coût par unité de longueur, $\frac{d}{dn}$ représente la dérivée suivant la normale à la courbe $\mu =$ constante, et φ est l'angle que fait la trajectoire du déplacement avec cette normale. Quand μ croît, la trajectoire s'incurve vers la normale, et sinon, elle s'en éloigne.

Les transformations conformes permettent dans un certain nombre de cas de simplifier le calcul des itinéraires à meilleur marché, ainsi que les courbes de même coût établies pour une même origine des déplacements. Dans les villes à symétrie circulaire on peut utiliser les transformations du type $\omega = z^p$ auxquelles correspond $\mu = cr^{p-1}$ avec r représentant la distance par rapport au centre de la ville. Le cas particulier $\omega = \log z$ correspond à $\mu = c \cdot r^{-1}$.

Cette étude apporte des exemples mis au point en fonction de temps de parcours minimum pour les cas où la vitesse est proportionnelle à $r^{1/2}$ ou à r.

KURZFASSUNG

Strecken mit minimalen Kosten in Stadtregionen

In einer früheren Abhandlung wurde gezeigt, daß eine direkte Analogie zwischen dem Verlauf von Strecken mit minimalen Transportkosten und den Bahnen von Lichtstrahlen besteht. Dieser Analogie folgend wird eine Gleichung für eine Route mit minimalen Kosten abgeleitet:

$$\varkappa = \frac{1}{\mu}\frac{d\mu}{dn}\sin\varphi$$

Dabei gibt $\varkappa$ die Krümmung an, μ sind die Kosten pro Streckeneinheit, $\frac{d}{dn}$ entspricht dem Differential entlang der Normalen für die Kurve $\mu =$ konstant, und φ ist der Winkel zwischen der Bahn und der Normalen. Wenn μ wächst, nähert sich die Bahn der Normalen; nimmt μ ab, so entfernt sie sich von der Normalen.

Konforme Transformationen können in verschiedenen Fällen angewendet werden, um die Berechnung von Routen mit minimalen Kosten und von Kurven gleicher Kosten von einem gegebenen Ursprung aus zu vereinfachen. Für radial symmetrische Städte können Transformationen der Form $\omega = z^p$ benützt werden, entsprechend $\mu = cr^{p-1}$, wobei r die Entfernung vom Stadtzentrum bedeutet. Der Sonderfall $\omega = \log z$ entspricht dem Fall $\mu = c \cdot r^{-1}$.

In Beispielen werden die minimalen Reisezeiten für die Fälle dargestellt, in denen die Geschwindigkeit proportional zu $r^{1/2}$ oder zu r ist.

INTRODUCTION

There are many problems in transport which are concerned with finding the path of minimum cost between given points. Examples are the choice of route to a given destination on a network of roads to minimise length or journey time, the choice of form of transport to minimise an individual's total outlay and the choice of a route for a new road.

The costs which are considered may be only operating

costs (journey time being a simple example), only construction costs or a combination of the two. They may also take account of a number of other factors, such as safety, comfort, convenience and amenity. The type of cost which we shall consider here is fairly general in the first instance. It will be assumed that the cost of transport is a function of location alone, not of direction. This may apply as a reasonable approximation to travel in a given transport system and to the building of new roads or railways, although in real life the costs may also depend on the direction of travel.

REFRACTION OF TRAFFIC

It has been pointed out by PALANDER [1a], VON STACKELBERG [1b] and LÖSCH [1c] that there is a direct analogy between minimum-cost paths on a plane in which cost per unit distance varies with location and the paths of light rays. They showed that Snell's Law (see below) applies to the path of minimum cost between two points in two regions of differing cost per unit distance. For the sake of completeness a proof of Snell's Law will be given here.

REFRACTION ACROSS A BOUNDARY

Consider two regions separated by a linear boundary, the cost of travel per unit length in any direction being μ in the first region and ν in the second (see Fig. 1). Let P be a point in the first region and R a point in the second. We wish to find the cheapest route (i. e. the route of minimum cost) joining P to R.

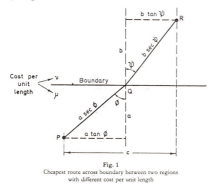

Fig. 1
Cheapest route across boundary between two regions
with different cost per unit length

Let Q be any point on the boundary. The shortest distance from P to Q is the straight line PQ and, since this lies in a region of uniform cost, this is also the cheapest route from P to Q. Similarly, the cheapest route from Q to R is the straight line QR. The total cost of the cheapest route via Q is therefore given by

$$x = \mu \cdot PQ + \nu \cdot QR$$
$$= \mu \cdot a \cdot \sec\varphi + \nu \cdot b \cdot \sec\psi \qquad (1)$$

where a, b, φ and ψ are defined in Fig. 1.

We wish to find the position of Q which minimises x. Now the diagram shows that

$$a \cdot \tan\varphi + b \cdot \tan\psi = c. \qquad (2)$$

a, b and c being constants. The minimum will occur when

$$\frac{dx}{d\varphi} = 0, \text{ i. e. when}$$

$$\mu \cdot a \cdot \sec\varphi \tan\varphi + \nu \cdot b \cdot \sec\varphi \tan\varphi \frac{d\psi}{d\varphi} = 0. \qquad (3)$$

But from (2)

$$a \cdot \sec^2\varphi + b \cdot \sec^2\psi \frac{d\psi}{d\varphi} = 0$$

and substituting for $\dfrac{d\psi}{d\varphi}$ in (3) gives

$$\mu \cdot a \cdot \sec\varphi \tan\varphi - \nu \cdot a \cdot \sec^2\varphi \sin\psi = 0$$

or

$$\mu \cdot \sin\varphi = \nu \cdot \sin\psi. \qquad (4)$$

This is Snell's Law of refraction, where μ and ν are the refractive indices of the two media. Since the refractive index is inversely proportional to the speed of light this shows that light rays follow paths of minimum journey time across a boundary.

Strictly, we should verify that this law gives a minimum journey time, not merely a stationary one. This can be done by differentiating the expression for x again; the differential coefficient will be found to be positive when $\mu \cdot \sin\varphi = \nu \cdot \sin\psi$.

REFRACTION IN A MEDIUM OF CONTINUOUSLY VARYING COST PER UNIT DISTANCE

Suppose that we wish to find the equation of the minimum-cost path in a continuously varying medium. Problems of this type have been discussed by BECKMANN [2]. We can apply Snell's Law to a series of parallel layers of infinitesimal thickness (see, for example, HEATH [3]). Let these layers be as shown in Fig. 2. We assume that a particular layer has thickness dn and cost per unit length μ. The cost is assumed to be $\mu + d\mu$ in the next layer.

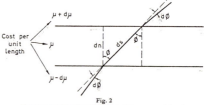

Fig. 2
Cheapest route through region with variable cost per unit length

Applying Snell's Law to the boundary between these two layers gives

$$\mu \cdot \sin\varphi = (\mu + d\mu) \sin(\varphi + d\varphi)$$
$$= \mu \cdot \sin\varphi + \mu \cdot \cos\varphi \, d\varphi + d\mu \cdot \sin\varphi$$

to the first order or

$$\mu \cdot \cos\varphi \, d\varphi = - \, d\mu \cdot \sin\varphi. \qquad (5)$$

For a general result we need to express this in terms of the curvature of the path, which is given by

$$\varkappa = - \frac{d\varphi}{ds}$$

with the usual sign convention. Now $ds = dn \cdot \sec\varphi$ (see Fig. 2). Hence

Wardrop

$$\varkappa = -\frac{d\varphi}{dn}\cos\varphi$$

$$= -\frac{d\varphi}{d\mu}\frac{d\mu}{dn}\cos\varphi$$

i. e.

$$\varkappa = \frac{1}{\mu}\cdot\frac{d\mu}{dn}\sin\varphi. \qquad (6)$$

In this equation $d\mu/dn$ is the rate of change of μ in the direction of the normal to the curve $\mu = $ constant.

The direction of the curvature is such that the path curves towards the normal if μ is increasing along the path, and away from it if μ is decreasing.

Equation (6) can be used to calculate the equations of cheapest routes in a plane in which the cost per unit distance is a given function of position, either by analytical integration or by numerical methods.

CIRCULAR CITIES

A great deal of attention has been paid to the analysis of circular cities or city centres (see, for example, SMEED [4] HOLROYD [5], and LAM and NEWELL [6]). LAM and NEWELL in particular considered the effects of congestion on routeing in a circular city with a ring and radial road system.

If direct paths are followed in a circular city, the traffic intensity builds up towards the centre. SMEED [7] calculated the traffic intensity at different distances from the centre for traffic entering a circular area in which destinations are uniformly distributed, and HOLROYD and MILLER [8] gave the corresponding intensity for internal journeys between random pairs of points. The results which they gave are shown in Table 1 in arbitrary units.

Table I

Traffic intensity in a circular city

Distance from centre divided by radius of city	Relative traffic intensity	
	Traffic entering, random destinations	Internal traffic between random points
0.0	1.00	1.00
0.2	0.96	0.95
0.4	0.90	0.80
0.6	0.80	0.57
0.8	0.67	0.29
1.0	0.42	0.00

On most road systems increases in traffic intensity produce increases in journey time per mile. WARDROP [9] has shown that on roads in the central areas of towns there is an approximate relation between journey speed and flow which can be put in the form

$$v = a\sqrt{(1-x)} \qquad (7)$$

where v is the journey speed, a is a constant and x is the ratio of the flow to the capacity. (The relation only applies if $x < 1$.) From this equation we derive Table 2.

Table 2 shows that as x approaches unity, i. e. as the flow nears the capacity, journey time per unit distance varies very strongly with flow.

The capacity of the road system tends to increase as the centre is approached. OWENS [10] has shown that, in a

Table II

Effects of flow on journey speed and journey time in a central area (arbitrary units)

Ratio of flow to capacity (λ)	Journey speed (v)	Journey time per unit distance (1/v)
0.0	1.00	1.0
0.2	0.89	1.1
0.4	0.77	1.3
0.6	0.63	1.6
0.8	0.45	2.2
0.9	0.32	3.2
0.95	0.22	4.5
0.97	0.17	5.8
0.98	0.14	7.1
0.99	0.10	10.0

sample of three towns in England, the proportion of ground space devoted to roads is approximately proportional to e^{-cd} where d is the distance from the town centre and c is a constant, ranging from 0.47 to 0.56 when d is measured in kilometres.

Despite the tendency of capacity per unit area to be less further away from the centre, it is highly probable that if direct routes are followed, the journey time per unit distance will decrease as distance from the centre increases. If drivers choose direct routes* initially, those who seek quickest routes will tend to avoid the city centre, thus lessening the traffic intensity near the centre. By repeated re-adjustment an equilibrium will eventually be reached in which no further reduction in journey time between any origin-destination pair is possible. In this condition the journey time per unit distance will still vary with distance from the centre. It seems, therefore, worth while exploring situations of this type. (For an example, see APPENDIX.)

CONFORMAL TRANSFORMATION

Conformal transformations using complex variables are very useful in many branches of physics, and they have an application in transport. It is easy to show that in certain circumstances a conformal transformation will change minimum-cost curves into straight lines. Suppose that z and ω are complex variables and $f(z)$ is a function which is differentiable almost everywhere. Consider the transformation

$$\omega = f(z). \qquad (8)$$

Let

$$\mu = \left|\frac{d\omega}{dz}\right|. \qquad (9)$$

Then if P is a path in the z-plane on which $\int_P \mu\, dz$ is a minimum, this transforms into a straight line in the ω-plane. For

$$\int_P \mu\,|dz| = \int_P \left|\frac{d\omega}{dz}\right|\,|dz| = \int_Q |d\omega|$$

where Q is the corresponding path in the ω-plane. But $\int_Q |d\omega|$ is the length of Q and if this is to be a minimum, Q must be a straight line. This property allows us to find minimum-cost paths easily in a variety of particular cases, in which μ satisfies an equation of the form (9). Some of these will now be examined.

* In practice, shortest routes in distance using the available road network.

CIRCULAR SYMMETRY

Consider the tranformation

$$\omega = z^p \tag{10}$$

where p is a real non-zero number. Then

$$\frac{d\omega}{dz} = p\, z^{p-1}$$

If we write

$$z = r\, e^{i\theta}$$

we have

$$\frac{d\omega}{dz} = p\, r^{p-1}\, e^{i(p-1)\,\theta}$$

and

$$\mu = \left|\frac{d\omega}{dz}\right| = |p|\, r^{p-1} \tag{11}$$

This represents a case in which the cost per unit distance varies with r (distance from the centre), and is independent of θ. If $p < 1$, μ decreases as r increases. According to the value of p the relation between μ and r is weaker or stronger.

We are often interested in minimum-cost paths from a given origin, and in the curves of equal cost from this point (sometimes called iso-cost curves). In what follows, to simplify the presentation, the costs will be taken as journey times. Consider quickest routes starting from $z = z_0$. The corresponding point in the ω-plane is

$$\omega_0 = z_0^p . \tag{12}$$

In this plane quickest routes are straight lines through ω_0, which we may represent parametrically as

$$\omega = \omega_0 + t\, e^{i\alpha} \tag{13}$$

Here α is constant for a particular route, and it is proportional to journey time along it. (In the ω-plane cost – in this case journey time – per unit distance is uniform. We have taken it as unity.) If t is fixed and α is allowed to vary, we have the curve of constant journey time t, which is, of course, a circle with radius t and centre ω_0.

Let us write

$$\omega = Re^{i\varphi} .$$

We may suppose that z_0 is real, since the curves have circular symmetry. It follows that ω_0 is also real. Then

$$Re^{i\varphi} = (\omega_0 + t \cdot \cos\alpha) + i \cdot t \cdot \sin\alpha$$

and so

$$R^2 = (\omega_0 + t \cdot \cos\alpha)^2 + (t \cdot \sin\alpha)^2$$

or

$$R^2 = \omega_0^2 + 2\omega_0 \cdot t \cdot \cos\alpha + t^2 \tag{14}$$

and

$$\cot\varphi = \frac{\omega_0}{t}\csc\alpha + \cot\alpha. \tag{15}$$

The point in the z-plane corresponding to ω is given by

$$z = \omega^{(1/p)} = R^{(1/p)}\, e^{i\,(\varphi/p)} .$$

Hence

$$r = R^{(1/p)} \tag{16}$$

$$\theta = \varphi/p. \tag{17}$$

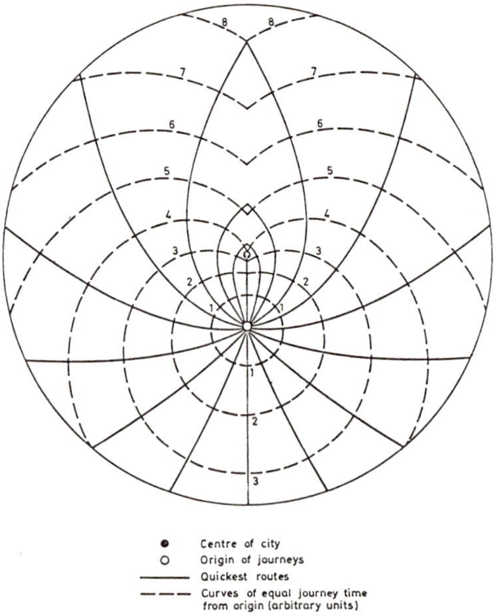

- ● Centre of city
- ○ Origin of journeys
- —— Quickest routes
- - - - Curves of equal journey time from origin (arbitrary units)

Fig. 3
Quickest routes in a city in which speed is proportional to
(distance from centre)$^{1/4}$

Wardrop

By successive application of equations (12), (14) to (17) we can calculate the quickest routes and the curves of equal journey time from the origin in the original plane (the z-plane).

Example I: $p = \dfrac{1}{2}$

If we take $p = \dfrac{1}{2}$ we have

$$\omega = z^{1/2}$$
$$\mu = cr^{-1/2} \quad \text{(c constant)}$$

and if v is the journey speed,

$$v = \mu^{-1}$$
$$= c'r^{1/2} \quad \text{(c' constant)}.$$

Whatever the value of p, we may take $z_o = 1$ without loss of generality because variations in z_o, and hence in ω_o, only have the effect of changing the scale of r and t by a constant factor (see equations (14) and (15)).

In this case it is convenient to write $z = x+iy$ (x, y real). Equations (13) becomes

$$\omega = 1 + te^{i\alpha}$$

and

$$z = \omega^2$$
$$= 1 + 2te^{i\alpha} + t^2 e^{2i\alpha}.$$

Hence

$$x = 1 + 2t \cdot \cos\alpha + t^2 \cdot \cos 2\alpha$$
$$y = 2t \cdot \sin\alpha + t^2 \cdot \sin 2\alpha.$$

Curves given by keeping α fixed and varying t (the quickest routes), and those given by keeping t fixed and varying α (the curves of equal journey time from the origin of the journeys) are plotted in Fig. 3. In order to display the symmetry of the families of curves the axes have been rotated through 90°. The arbitrary journey time units correspond to increments of $\frac{1}{4}$ in the value of t.

It will be seen that the quickest routes are curved around the centre of the city and thus avoid it. The quickest route to the center itself is along the radius, however. All points beyond the centre are reached by routes avoiding the centre. The curves of equal journey time are more or less circular near the origin of journeys but are distorted when they pass near the centre. The 4-unit curve is a cardioid with the cusp at the centre.

The diagram shows that – as would be expected – points equidistant from the origin on the radius towards the centre are appreciably further in time than those at the same distance on the outwards radius, the ratios of the times over much of the range being between 1.5 and 2.5.

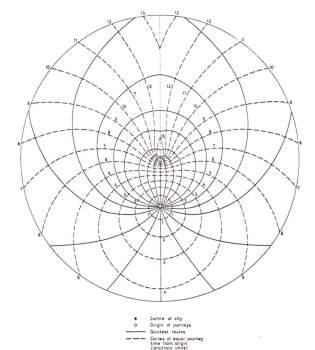

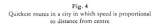

•	Centre of city
o	Origin of journeys
——	Quickest routes
– – –	Curves of equal journey time from origin (arbitrary units)

Fig. 4
Quickest routes in a city in which speed is proportional to distance from centre

Example II: $\mu = cr^{-1}$

This special case is not covered by the previous analysis. If, however, we write

$$\frac{d\omega}{dz} = z^{-1}$$

we have the required law

$$\mu = cr^{-1} \qquad \text{(c constant)}$$

or

$$v = c'r \qquad \text{(c' constant)}.$$

Integrating gives

$$\omega = \log z$$

or

$$z = e^{\omega}.$$

As before we can take $z = 1$ as origin of journeys without loss of generality. In this case the (a, t) points in the ω-plane are given by

$$\omega = te^{ia}$$

so

$$z = e^{t \cdot \cos a} \, e^{i \cdot t \cdot \sin a}$$

i.e.

$$r = e^{t \cdot \cos a}$$
$$\theta = t \cdot \sin a.$$

The quickest routes (allowing t to vary but keeping a fixed) are given by

$$dr = e^{t \cdot \cos a} \cos a \, dt$$
$$= r \cdot \cos a \, dt$$
$$d\theta = \sin a \, dt.$$

Hence

$$\frac{r d\theta}{dr} = \tan a.$$

This shows that a quickest route curve makes a constant angle with the radius from the origin, i. e. it is an equiangular spiral. (MILLER [11] has discussed the properties of road systems consisting of equiangular spirals and circular arcs.)

The corresponding curves in this case are plotted in Fig. 4. Almost all the quickest routes show a pronounced curvature round the centre of the city. A particular example of a quickest route is a circle with centre at the city centre. If the origin were at the edge of the city this circle would also be the perimeter of the city.

APPLICATIONS

Models of the type considered could be regarded as approximations to real city centres, and calculations made of such quantities as the average distance travelled, the average direct speed (straight-line distance divided by journey time) and the traffic intensity as a function of distance from the centre. It is easy to find the quickest route between any two specified points, so that by choosing random pairs of points the distributions of various characteristics of the traffic could be readily estimated.

OTHER CASES

There are no other conformal tranformations which preserve circular symmetry. If a law of the form $\mu = cr^p$ is not appropriate, the city could be divided into annular regions in each of which a law of this form applies. In such cases, of course, it would be necessary to plot separate sets of curves for origins at different distances from the centre and the process would become complex.

Transformations can be found in which speed is reduced as one approaches each of several centres. For example, if

$$\omega = (z-1)^{1/2} + (z+1)^{1/2}$$
$$\frac{d\omega}{dz} = \frac{1}{2} (z-1)^{-1/2} + \frac{1}{2} (z+1)^{-1/2}$$

and $\mu = \left|\frac{d\omega}{dz}\right|$ approaches infinity, i. e. v approaches zero,

near $z = 1$ and $z = -1$. More generally by taking

$$\omega = (z-a_1)^{p_1} + (z-a_2)^{p_2} + \ldots + (z-a_n)^{p_n}$$

where $p_i < 1$, $j = 1 \ldots n$, it would be possible to simulate an area with a number of centres of congestion, with different "weights".

CONCLUSION

It appears that where the cost per unit distance (or speed) varies continuously throughout an area, conformal transformations can often be used to simplify the process of determining cheapest (or quickest) routes. This process also allows iso-cost curves (or curves of equal journey time) to be plotted easily, and could be used to study the properties of a variety of transport systems.

ACKNOWLEDGEMENTS

The author is grateful to the Construction Industry Research and Information Association for support for this investigation.

REFERENCES

[1a] PALANDER, T.:
Beiträge zur Standorttheorie, Uppsala, 1935.

[1b] von STACKELBERG, H.:
Das Berechnungsgesetz des Verkehrs, Jahrbücher für Nationalökonomie und Statistik, 148, 680–696, 1938.

[1c] LÖSCH, A.:
The economics of location, Science Editions, John Wiley & Sons, Inc., New York, 1967, pages 184–7, (Die räumliche Ordnung der Wirtschaft. G. Fischer-Verlag, Jena, 1940).

[2] BECKMANN, M. J.:
A continuous model of transportation, Econometrica, 20, 643–66, 1952.

[3] HEATH, R. S.:
A treatise on geometrical optics, pages 356–369, Cambridge University Press, 1895.

[4] SMEED, R. J.:
A theoretical model of commuter traffic in towns, 1. Inst. Math. Applics, 1 (3), 208–225, 1965.

[5] HOLROYD, E. M.:
Theoretical average journey lengths in circular towns with various routeing systems, Ministry of Transport, RRL Report No. 43. Crowthorne, 1966.

[6] LAM, T. N. and NEWELL, G. F.:
Flow dependent traffic assignment on a circular city, Transportation Science, 1 (4), 1967.

[7] SMEED, R. J.:
The traffic problem in towns, (Manchester Statistical Society) Manchester, 1961.

Wardrop — Potts and Loubal

[8] HOLROYD, E. M. and MILLER, A. J.:
Route crossings in urban areas, Proc. Austr. Rd Res. Bd. 3 (1),394–419, 1966.

[9] WARDROP, J. G.:
Journey speed and flow in central urban areas, Traffic Engng & Control, 9 (11), 528–32, 539, 1966.

[10] OWENS, D.:
Estimates of the proportion of space occupied by roads and footpaths in towns, Ministry of Transport RRL Report No. LR. 154, Crowthorne, 1968.

[11] MILLER, A. J.:
On spiral road networks, Transportation Science, 1 (2), 109–25, 1967.

APPENDIX

Journey Times and Speeds in London. In London the average times taken to travel one mile by car on main roads in off-peak hours at various distances from the centre (Charing Cross), and the corresponding speeds are as follows (s. Table):

These results were derived from data given in the London Traffic Survey, Vol. I, 1964 (London County Council)

Distance from centre (miles)	Journey time (minutes per radial mile)	Speed (radial miles per hour)
0 – 2	6.00	10.0
2 – 4	4.40	13.6
4 – 7	3.27	18.3
7 – 11	2.62	22.9
11 – 15	2.25	26.7

Reprinted from *Transportation Research*, 1, 253–269, 1967

Transpn Res. Vol. 1, pp. 253–269. Pergamon Press 1967. Printed in Great Britain

A STATISTICAL THEORY OF SPATIAL DISTRIBUTION MODELS

A. G. WILSON

Economic Planning Group, Ministry of Transport, London, England

(*Received* 9 *May* 1967)

1. INTRODUCTION

THIS paper offers a theoretical basis for some spatial distribution models which are in common use in locational analysis, and uses the new methodology to generate new models. Such analysis is likely to be concerned with the spatial location of activities among the zones of a region, and measures of interaction between the zones. For example, there may be an interest in zonal activity levels in the form of numbers of workers in residential zones, and the numbers of jobs in employment zones; the interaction between these activities is the journey to work. To fix ideas, this example will be used to illustrate the methods proposed in this paper.

Let T_{ij} be the number of (work) trips, and d_{ij} the distance, between zones i and j, let O_i be the total number of work trip origins in i, and let D_j be the total number of work trip destinations in j. A spatial distribution model then estimates T_{ij} as a function of the O_i's, the D_j's and the d_{ij}'s. These variables could, of course, themselves be functions of other independent variables.

The simplest such model is the so-called gravity model developed by analogy with Newton's law of the gravitational force F_{ij} between two masses m_i and m_j separated by a distance d_{ij}.

$$F_{ij} = \gamma \frac{m_i m_j}{d_{ij}^2} \qquad (1)$$

where γ is a constant. The analogous transport gravity model is then

$$T_{ij} = k \frac{O_i D_j}{d_{ij}^2} \qquad (2)$$

using the variables defined above, and where k is a constant. This model has some sensible properties: T_{ij} is proportional to each of O_i and D_j and inversely proportional to the square of the distance between them. But the equation has at least one obvious deficiency: if a particular O_i and a particular D_j are each doubled, then the number of trips between these zones would quadruple according to the equation (2), when it would be expected that they would double also. To put this criticism of (2) more precisely, the following constraint equations on T_{ij} should always be satisfied, and they are not satisfied by (2):

$$\sum_j T_{ij} = O_i \qquad (3)$$

$$\sum_i T_{ij} = D_j \qquad (4)$$

That is, the row and column sums of the trip matrix should be the numbers of trips generated in each zone, and the number of trips attracted, respectively. These constraint equations

18   253

can be satisfied if sets of constants A_i and B_j associated with production zones and attraction zones respectively are introduced. They are sometimes called balancing factors. Also, there is no reason to think that distance plays its part in the transport equation (2) as it does in the world of Newtonian physics, and so a general function of distance is introduced. The modified gravity model is then

$$T_{ij} = A_i B_j O_i D_j f(d_{ij}) \tag{5}$$

where

$$A_i = \left[\sum_j B_j D_j f(d_{ij}) \right]^{-1} \tag{6}$$

and

$$B_j = \left[\sum_i A_i O_i f(d_{ij}) \right]^{-1} \tag{7}$$

The equations for A_i and B_j are solved iteratively, and it can easily be checked that they ensure that the T_{ij} given in equation (5) satisfies the constraint equations (3) and (4). Note also that d_{ij} in such a model should be interpreted as a general measure of impedance between i and j, which may be measured as actual distance, as travel time, as cost, or as some weighted combination of such factors sometimes referred to as a "generalized cost". With this proviso, equations (5)–(7) describe a gravity model which has been extensively used, and the discussion above has shown its heuristic derivation by analogy with Newton's gravitational law.

A second approach to trip distribution uses the intervening opportunities model. Interzonal impedance does not appear explicitly in this model, but possible destination zones away from an origin zone i have to be ranked in order of increasing impedance from i. A notation is needed to describe this. Let $j_\mu(i)$ be the μth destination zone in this rank order away from i; $j_\mu(i)$ will be referred to simply as j_μ in cases where it is clear to which i it refers. The intervening opportunities model was first developed by Stouffler (1940) in a simple form, assuming that the number of trips from an origin zone to a destination zone is proportional to the number of opportunities at the destination zone, and inversely proportional to the number of intervening opportunities. The underlying assumption of the model is that the tripper considers each opportunity, as reached, in turn, and has a definite probability that his needs will be satisfied. The model will be derived here in the form developed by Schneider, originally for use in the Chicago Area Transportation Study (1960). To see how the basic assumption operates, consider a situation in which destination zones are rank ordered away from an origin zone as defined earlier. Let U_{ij_μ} be the probability that one tripper will continue beyond the μth zone away from i. Suppose there is a chance L that an opportunity will satisfy this single tripper when it is offered. Then, to the first order in L,

$$U_{ij_1} = 1 - L D_{j_1}$$

where D_{j_1} is the number of opportunities in the zone j_1, nearest to i. Then combining successive probabilities multiplicatively,

$$U_{ij_2} = U_{ij_1}(1 - L D_{j_2})$$
$$U_{ij_3} = U_{ij_2}(1 - L D_{j_3})$$

and so on. In general

$$U_{ij_\mu} = U_{ij_{\mu-1}}(1 - L D_{j_\mu}) \tag{8}$$

This equation can be written

$$\frac{U_{ij_\mu} - U_{ij_{\mu-1}}}{U_{ij_{\mu-1}}} = -L D_{j_\mu}. \tag{9}$$

A statistical theory of spatial distribution models   255

Let A_{j_μ} be the number of opportunities passed up to and including zone j_μ. Then

$$D_{j_\mu} = A_{j_\mu} - A_{j_{\mu-1}} \tag{10}$$

and (9) can be written

$$\frac{U_{ij_\mu} - U_{ij_{\mu-1}}}{U_{ij_{\mu-1}}} = -L(A_{j_\mu} - A_{j_{\mu-1}}) \tag{11}$$

This equation can be written, assuming continuous variation, as

$$\mathrm{d}U/U = -L\,\mathrm{d}A \tag{12}$$

which integrates to

$$\log U = -LA + \text{constant}$$

so that

$$U_{ij_\mu} = k_i \exp(-LA_{j_\mu}) \tag{13}$$

where k_i is a constant. But

$$T_{ij_\mu} = O_i(U_{ij_{\mu-1}} - U_{ij_\mu}) \tag{14}$$

where T_{ij_μ} is the number of trips from i to the μth destination away from i, for a total of O_i trips originating at i. Substitution from (13) to (14) gives

$$T_{ij_\mu} = k_i O_i [\exp(-LA_{j_{\mu-1}}) - \exp(-LA_{j_\mu})] \tag{15}$$

and this is the usual statement of the intervening opportunities model.

Note that k_i can be chosen so that the resulting matrix T_{ij} satisfies the constraint equation (3):

$$\sum_j T_{ij} = k_i O_i [1 - \exp(-A_{j_N})] = O_i$$

where N is the total number of zones. Since $\exp(-LA_{j_N})$ should be very small, k_i will be very nearly one for each i. The constraint equation (4) on the total number of trip attractions cannot be satisfied, however, within the model structure itself, but if actual D_j's are known, the matrix can be adjusted by the same balancing process as that implied by equations (6) and (7).

(Thus, in general, if there is a matrix $T_{ij}{}^*$ and it is required to transform it to a matrix T_{ij} whose columns and rows sum to O_i and D_j, then this can be done by the transformation

$$T_{ij} = A_i B_j T_{ij}{}^* \tag{16}$$

where

$$A_i = O_i \left(\sum_j B_j T_{ij}{}^* \right)^{-1} \tag{17}$$

$$B_j = D_j \left(\sum_i A_i T_{ij}{}^* \right)^{-1} \tag{18}$$

and these equations can be seen to reduce to (6) and (7) for the gravity model. This process is accomplished in practice by factoring rows and columns successively by $D_j/D_j{}^*$ and $O_i/O_i{}^*$ where the $D_j{}^*$ are the column sums, and $O_i{}^*$ the row sums, of the matrix reached after the immediately preceding operation. This is the balancing process which can be applied, if required, to the intervening opportunities model trip matrix.)

It is often argued that the intervening opportunities model is "better" than the gravity model because the theoretical derivation (as outlined above) is sound, whereas the derivation of the gravity model is at best heuristic and based on an analogy with Newton's gravitational law in the physical sciences. The statistical theory of spatial distribution models proposed in this paper is based on an analogy with a different branch of physics, statistical mechanics, and does offer a sound theoretical base for the gravity model. It is also possible, by changing the assumptions, to derive the intervening opportunities model. In fact, the proposed method provides a reasonably general rule for deriving appropriate spatial distribution models for a variety of purposes and situations, and for comparing different models which are supposed to apply to the same situation.

The key to the method is to define a set of variables which completely specify the system, and to enumerate any constraints on these variables. It will usually be possible to do this using the variables T_{ij} defined previously, the number of trips between i and j (still assuming that this is for one purpose and a reasonably homogeneous set of travellers). A set of T_{ij}'s, defined by $\{T_{ij}\}$ then defines a *distribution* of trips. It is also possible to define a *state* of the system as one of the ways in which a distribution is brought about at the micro-level: thus, if the system is made up of individual trippers, a state of the system, for example for the journey to work, is just one way in which these trippers decide on their journey to work in a consistent way (that is, subject to the usual constraints). Note that, at the end of the analysis, the only interest is in distributions—total numbers of trips between points— and not in states—that is, the individual trippers who are making up these trip bundles. The crucial assumption of the new method can now be stated: that the probability of a distribution $\{T_{ij}\}$ occurring is proportional to the number of states of the system which give rise to the distribution $\{T_{ij}\}$. Thus, if $w(T_{ij})$ is the number of ways in which individuals can arrange themselves to produce the overall distribution $\{T_{ij}\}$, then the probability of $\{T_{ij}\}$ occurring is proportional to $w(T_{ij})$. The total number of such arrangements is

$$\sum w(T_{ij}) \tag{19}$$

where the summation is over the distributions which satisfy the constraints of the problem. It will be shown that in most cases there is one distribution $\{T_{ij}\}$ for which $w(T_{ij})$ dominates all other terms of the sum (19) overwhelmingly, and so forms the most probable distribution.

Section 2 of this paper states and discusses this theory for the usual gravity model; Section 3 then applies the theory to a variety of new situations, including an application to multimodal distribution, and also including a derivation of the intervening opportunities model. The final section summarizes the conclusions.

2. A STATISTICAL THEORY OF THE GRAVITY MODEL

2.1. The "conventional" gravity model

The new method is most easily illustrated by a single-purpose example, such as the journey to work example discussed above, and so the previously defined example and notation are used. The trip matrix, T_{ij}, must satisfy the two constraint equations (3) and (4), which, for convenience, are stated again here:

$$\sum_j T_{ij} = O_i \tag{20}$$

$$\sum_i T_{ij} = D_j \tag{21}$$

It will also be assumed that another constraint equation is satisfied:

$$\sum_i \sum_j T_{ij} c_{ij} = C \tag{22}$$

where c_{ij} is the impedance, or generalized cost, of travelling between i and j, and so replaces the d_{ij} of the introduction to emphasize that the measure of impedance need not be distance.

A statistical theory of spatial distribution models   257

This constraint, then, implies that the total amount spent on trips in the region at this point in time, is a fixed amount C. The use of, and need for, this constraint will be made clear as the method develops.

The basic assumption of the method is that the probability of the distribution $\{T_{ij}\}$ occurring is proportional to the number of states of the system which give rise to this distribution, and which satisfy the constraints. Suppose

$$T = \sum_i O_i = \sum_j D_j \tag{23}$$

is the total number of trips. Then the number of distinct arrangements of individuals which give rise to the distribution $\{T_{ij}\}$ is

$$w(T_{ij}) = (T!) \Big/ \left(\prod_{ij} T_{ij}!\right) \tag{24}$$

since there is no interest in arrangements within a particular trip bundle. The total number of possible states is then

$$W = \sum w(T_{ij}) \tag{25}$$

where the summation is over T_{ij} satisfying (20)–(22). However, the maximum value of $w(T_{ij})$ turns out to dominate the other terms of the sum to such an extent that the distribution $\{T_{ij}\}$, which gives rise to this maximum is overwhelmingly the most probable distribution. This maximum will now be obtained, and its sharpness, and the validity of the method in general, will then be discussed in Section 2.3 below, following a section which discusses the interpretation of particular terms.

To obtain the set of T_{ij}'s which maximizes $w(T_{ij})$ as defined in (24) subject to the constraints (20)–(22), the function M has to be maximized, where

$$M = \log w + \sum_i \lambda_i^{(1)} \left(O_i - \sum_j T_{ij}\right) + \sum_j \lambda_j^{(2)} \left(D_j - \sum_i T_{ij}\right) + \beta \left(C - \sum_i \sum_j T_{ij} c_{ij}\right) \tag{26}$$

and where $\lambda_i^{(1)}$, $\lambda_j^{(2)}$ and β are Lagrangian multipliers. Note that it is more convenient to maximize $\log w$ rather than w, and then it is possible to use Stirling's approximation

$$\log N! = N \log N - N \tag{27}$$

to estimate the factorial terms. The T_{ij}'s which maximize M, and which therefore constitute the most probable distribution of trips, are the solutions of

$$\frac{\partial M}{\partial T_{ij}} = 0 \tag{28}$$

and the constraint equations (20)–(22). Using Stirling's approximation, (27), note that

$$\frac{\partial \log N!}{\partial N} = \log N \tag{29}$$

and so

$$\frac{\partial M}{\partial T_{ij}} = -\log T_{ij} - \lambda_i^{(1)} - \lambda_j^{(2)} - \beta c_{ij} \tag{30}$$

and this vanishes when

$$T_{ij} = \exp(-\lambda_i^{(1)} - \lambda_j^{(2)} - \beta c_{ij}) \tag{31}$$

Substitute in (20) and (21) to obtain $\lambda_i^{(1)}$ and $\lambda_j^{(2)}$:

$$\exp\left[-\lambda_i^{(1)}\right] = O_i \bigg/ \left[\sum_j \exp\left(-\lambda_j^{(2)} - \beta c_{ij}\right)\right] \tag{32}$$

$$\exp\left[-\lambda_j^{(2)}\right] = D_j \bigg/ \left[\sum_i \exp\left(-\lambda_i^{(1)} - \beta c_{ij}\right)\right] \tag{33}$$

To obtain the final result in more familiar form, write

$$A_i = \exp\left(-\lambda_i^{(1)}\right)/O_i \tag{34}$$

and

$$B_j = \exp\left(-\lambda_j^{(2)}\right)/D_j \tag{35}$$

and then

$$T_{ij} = A_i B_j O_i D_j \exp\left(-\beta c_{ij}\right) \tag{36}$$

where, using equations (32)–(35),

$$A_i = \left[\sum_j B_j D_j \exp\left(-\beta c_{ij}\right)\right]^{-1} \tag{37}$$

$$B_j = \left[\sum_i A_i O_i \exp\left(-\beta c_{ij}\right)\right]^{-1} \tag{38}$$

Thus the most probable distribution of trips is the same as the gravity model distribution discussed earlier, and defined in equations (5)–(7), and so this statistical derivation constitutes a new theoretical base for the gravity model. Note that C in the cost constraint equation (22) need not actually be known, as this equation is not in practice solved for β. This parameter would be found by the normal calibration methods. However, if C was known, then (22) could be solved numerically for β.

This statistical theory is effectively saying that, given total numbers of trip origins and destinations for each zone for a homogeneous person–trip purpose category, given the costs of travelling between each zone, and given that there is some fixed total expenditure on transport in the region, then there is a most probable distribution of trips between zones, and this distribution is the same as the one normally described as the gravity model distribution. Students of statistical mechanics will recognize the method as a variation of the micro-canonical ensemble method for analysing systems of particles, for example, the molecules of a gas.

2.2. Interpretation of terms

It has always been a feature of statistical mechanics that the terms which occur in the equation giving the most probable distribution are then seen to have physical significance. This is true here also. O_i, D_j and c_{ij} were defined previously. The expression $\exp\left(-\beta c_{ij}\right)$ appears in this formulation as the preferred form of distance deterrence function, and the parameter β is determined in theory by the cost constraint equation (22). It does, however, have its usual interpretation: it is closely related to the average distance travelled. The greater β, the less is the average distance travelled. This is then obviously related to C of equation (22). If C is increased, then more is spent on travelling and distances will increase, but an examination of the left-hand side of (22) shows that β would decrease. The remaining task is to interpret A_i and B_j.

Suppose one of the D_j's changes, say D_1. Then

$$T_{i1} = A_i B_1 O_i D_1 \exp\left(-\beta c_{i1}\right) \tag{39}$$

and if D_1 changes substantially, the trips from each i to zone 1 would change in proportion.

The next largest change will be in the A_i's as defined by (37), but the change will not be large as the expression involving D_1 in each A_i is only one of a number of terms. The B_j's will probably be affected even less, as any change is brought about through changes in the A_i's.

Suppose, then, that D_1 is substantially increased. Then the T_{i1}'s will increase more or less in proportion. The A_i's will decrease by a lesser amount relatively, and the B_j's will increase even more slightly. The role of the A_i's, then, will be to reduce all trips slightly to compensate for the increase in trips to zone 1. A_i can thus be seen as a competition term which reduces most trips due to the increased attractiveness of one zone. The denominator of A_i is also commonly used as a measure of accessibility, and the increase in D_1 could be said to increase the accessibility of everyone to opportunities at 1, though more usually such an interpretation would be reserved for changes in the c_{ij}'s. Thus, this analysis establishes a competition–accessibility interpretation of the A_i's. The B_j's play a similar role, and would be responsible for the main adjustments if the major change was in an O_i rather than a D_j. A change in the c_{ij}'s, or several O_i's and D_j's simultaneously, would bring about complex readjustments through the A_i's and the B_j's.

One consequence of this interpretation, and of the use of the new method which gives a fundamental role to the A_i's and B_j's, is that it shows that the interpretation of the A_i's and B_j's suggested by Dieter (1962) is wrong. Dieter suggested that the A_i's and B_j's should be associated with terminal costs, say a_i and b_j in origin and destination zones i and j respectively. This can be checked by replacing c_{ij} by $a_i + b_j + c_{ij}$ in the preceding analysis, and this gives for T_{ij}:

$$T_{ij} = A_i B_j O_i D_j \exp(-\beta a_i - \beta b_j - \beta c_{ij}) \qquad (40)$$

Thus, new terms $\exp(-\beta a_i)$ and $\exp(-\beta b_j)$ are introduced, but the A_i's and B_j's are still present independently of the existence of terminal costs.

2.3. Validity of the method

There are two possible points of weakness in the method. Firstly, is Stirling's approximation, in equation (27), valid for the sort of T_{ij}'s that occur in practice? Secondly, is the maximum value of

$$\frac{T!}{\prod_{ij} T_{ij}!}$$

a very sharp maximum?

The first of the doubts can be answered by analogy. The use of Stirling's approximation underlies one particular approach in statistical mechanics and is used, as here, to produce most probable distributions. There is, however, a second method, the Darwin–Fowler method, which actually calculates the individual terms of the sum in equation (19) by using a generating function and complex integration. These terms are then used as weights to calculate the means of all the distributions, and these mean values have been shown to be the same as the most probable values obtained by using Stirling's approximation, even in the cases where the numbers involved are obviously so small that Stirling's theorem is not valid. It is a safe conjecture that the same result applies here: that theoretically valid results can be obtained, using the method above, at all times.

The second question can be answered explicitly if small changes in $\log w(T_{ij})$ are examined near the maximum. At, or very near, the maximum the terms of $d[\log w(T_{ij})]$ which are linear in dT_{ij} vanish, and

$$d[\log w(T_{ij})] = \tfrac{1}{2} \sum_i \sum_j \frac{\partial^2 \log w}{\partial T_{ij}^2} (dT_{ij})^2 \qquad (41)$$

It will be recalled that

$$\frac{\partial \log w}{\partial T_{ij}} = -\log T_{ij}$$

and so

$$\frac{\partial^2 \log w}{\partial T_{ij}^2} = -T_{ij}^{-1} \tag{42}$$

Substituting in equation (41),

$$d\left[\log w(T_{ij})\right] = -\tfrac{1}{2}\sum_i \sum_j \frac{(dT_{ij})^2}{T_{ij}} = -\tfrac{1}{2}\sum_i \sum_j \left(\frac{dT_{ij}}{T_{ij}}\right)^2 T_{ij} \tag{43}$$

Thus (43) can be written

$$d\left[\log w(T_{ij})\right] = -\tfrac{1}{2}\sum_i \sum_j p^2 T_{ij} \tag{44}$$

where p is the percentage change in each T_{ij} away from the most probable distribution.

To evaluate this expression, the size distribution of elements of the trip matrix is needed. Suppose there are N size groups, and that the nth group has T_{ij}'s with a mean value T_n, and there are S_n such trip matrix elements in this group. Then, (44) can be written

$$d\left[\log w(T_{ij})\right] = -\tfrac{1}{2}p^2 \sum_n S_n T_n \tag{45}$$

Now consider a typical example: take a large urban area with, say 1000 zones. Suppose 1000 trip interchanges have each got 10^4 trips, 10,000 have 10^3 and 100,000 have 10^2. Let $p = 10^{-3}$. Then

$$d \log w \approx -\tfrac{1}{2}10^{-6}(10^7 + 10^7 + 10^7) \approx -15$$

Thus, $\log w$ changes by -15 for a change of one part in a thousand of each element of the trip matrix away from the most probable distribution. Thus w drops by the enormous factor of e^{-15}, which gives an indication of just how sharp the maximum can be. Such an estimate of $d(\log w)$ can be calculated as a check in any particular case. One of the advantages of this new approach is that it gives the possibility of doing this check, and ruling out certain situations as being unsuitable for the gravity model approach should the maximum turn out not to be a sharp one.

A third result of interest can also be stated by analogy with the corresponding result in statistical mechanics (cf. Tolman, 1938). That is:

$$\frac{\overline{(T_{ij} - \bar{T}_{ij})^2}}{\bar{T}_{ij}^2} = T_{ij}^{-1} - T^{-1} \tag{46}$$

This gives the dispersion of T_{ij} and indicates, as is well known in practice, that estimates are better for large flows than for small ones.

This analysis has shown that the gravity model has a sound base. However, it should be recalled that the whole analysis has been for a single trip purpose, and for a homogeneous group of travellers. People are not identical in the way that particles in physics are identical, and so no theory of this form (indeed, no theory period) can be expected to apply exactly. This analysis has shown, in effect, that good results can be expected if trips can be classified by purpose and by person type in a reasonably uniform way.

3. NEW APPLICATIONS OF THE STATISTICAL THEORY

3.1. *Rules for constructing distribution models*

The previous sections have shown that a statistical theoretical base can be given to the conventional gravity model. The principle on which this derivation is based is, however, quite general. The only assumption is that the probability of a distribution occurring is proportional to the number of states of the system which give rise to that distribution, subject to a number of constraints. It can easily be seen that if there were no constraints at all, then all the T_{ij}'s would have an equal share of the total number of trips. In other words, it is the constraints which have the effect of giving a distribution of trips other than a trivial one. One of the remarkable features of this statistical theory is that it produces the conventional gravity model using constraints which really say relatively little, that is, are relatively unrestrictive. The new theory reveals that an effective way of developing better models is to refine the constraints which are applied to behaviourable variables to make them more restrictive. This makes more precise for this problem what is simply the normal method of scientific research. (See, for example, Popper, 1959.) This new theory is a powerful general method for constructing spatial distribution models: the main task to produce a model in this framework is to discover the constraints on the variables which describe the spatial distribution problem, and then to maximize, as a function of the distribution variables, the number of states which can give rise to the distribution subject to the constraints.

The derivation of the conventional gravity model above followed this pattern. The following sections of the paper apply the general theory to a number of situations, ranging from new ones being tackled for the first time to a derivation of the intervening opportunities model.

3.2. The single competition term gravity model

The so-called conventional gravity model was described by equations (5)–(7) above. A much-used variant of this simply takes all the B_j's as one at the expense of failing to satisfy the constraints (4). So this model is

$$T_{ij} = A_i O_i D_j f(c_{ij}) \tag{47}$$

where

$$A_i = \left[\sum_j D_j f(c_{ij})\right]^{-1} \tag{48}$$

to ensure that

$$\sum_j T_{ij} = O_i \tag{49}$$

but where, if

$$D_j^* = \sum_i T_{ij},$$

D_j^* is not necessarily equal to D_j. This model may be used in a variety of special circumstances: for example, when D_j is simply some measure of attraction, and D_j^* is then *defined* to be the resulting number of trips produced in the model. That is, no constraint of the form (4) is assumed to hold. The new method can be applied to this situation easily: maximize

$$\frac{T!}{\prod_{ij} T_{ij}!}$$

subject to the constraints (49), and subject to a generalized cost constraint (22). The resulting distribution, which is analogous to (31) as derived from (30), is

$$T_{ij} = \exp(-\lambda_i^{(1)} - \beta c_{ij}) \tag{50}$$

where the $\lambda_i^{(1)}$'s and β are Lagrangian multipliers. The $\lambda_i^{(1)}$ can be found by substituting in (49) with the result

$$T_{ij} = \frac{O_i \exp(-\beta c_{ij})}{\sum_k \exp(-\beta c_{ik})} \tag{51}$$

Now this equation resembles (47) and (48), but the D_j term of the latter equation is now missing. However, it will be recalled that in this case D_j is likely to be a measure of attraction, and not a number of trip ends. Such a term can be introduced into equation (51) by the following interesting device. Assume that the traveller to j receives some benefit W_j which can be set against the transport cost c_{ij} over and above the benefit which can be obtained from going to other zones. (So W_j may be a measure of the scale economies available to the shopper in large shopping centres.) Equation (51) can then be rewritten with c_{ij} being replaced by $c_{ij} - W_j$ as

$$T_{ij} = \frac{O_i \exp(\beta W_j - \beta c_{ij})}{\sum_k \exp(\beta W_k - \beta c_{ik})} \tag{52}$$

so that we can now identify the new model with that of equations (47)–(48) by taking $\exp(\beta W_j)$ as the attractive measure D_j. Since D_j in such applications is usually taken as a zonal size variable which is a proxy for scale benefits of the form W_j, it seems intuitively reasonable to expect W_j to vary as $\log D_j$, as implied here, rather than D_j.

3.3. Distribution of trips when there are several transport modes

This section discusses a more important problem. Consider the basic situation illustrated earlier by the journey to work, and described by variables T_{ij}, O_i, D_j and c_{ij}. Suppose now, however, that there are several possible modes of transport between i and j, and the cost of travelling by the kth mode is c_{ij}^k. Let T_{ij}, O_i and D_j be defined as before, and let T_{ij}^k, O_i^k and D_j^k be the proportions of these trip totals carried by mode k.

The constraint equations which describe this situation are

$$\sum_i \sum_k T_{ij}^k = D_j \tag{53}$$

$$\sum_j \sum_k T_{ij}^k = O_i \tag{54}$$

$$\sum_i \sum_j \sum_k T_{ij}^k c_{ij}^k = C \tag{55}$$

and the maximand, subject to these constraints, is

$$\frac{T!}{\prod_{ijk} T_{ij}^k!}$$

Note that the constraint equations (53) and (54) do not use any knowledge of trip end modal split, and because of this, there are the same number of Lagrangian multipliers as for the single-mode case. The most probable distribution, obtained in the usual way, is

$$T_{ij}^k = \exp(-\lambda_i^{(1)} - \lambda_j^{(2)} - \beta c_{ij}^k) \tag{56}$$

and, substituting in (53) and (54),

$$\exp(-\lambda_j^{(2)}) = \frac{D_j}{\sum_i \sum_k \exp(-\lambda_i^{(1)} - \beta c_{ij}^k)} \tag{57}$$

$$\exp(-\lambda_i^{(1)}) = \frac{O_i}{\sum_j \sum_k \exp(-\lambda_j^{(2)} - \beta c_{ij}^k)} \tag{58}$$

So, putting

$$A_i = \exp(-\lambda_i^{(1)})/O_i, \quad B_j = \exp(-\lambda_j^{(2)})/D_j$$

as before,

$$T_{ij}{}^k = A_i B_j O_i D_j \exp(-\beta c_{ij}{}^k) \tag{59}$$

where

$$A_i = \left[\sum_j \sum_k B_j D_j \exp(-\beta c_{ij}{}^k) \right]^{-1} = \left[\sum_j B_j D_j \sum_k \exp(-\beta c_{ij}{}^k) \right]^{-1} \tag{60}$$

$$B_j = \left[\sum_i \sum_k A_i O_i \exp(-\beta c_{ij}{}^k) \right]^{-1} = \left[\sum_i A_i O_i \sum_k \exp(-\beta c_{ij}{}^k) \right]^{-1} \tag{61}$$

Note also that

$$T_{ij} = \sum_k T_{ij}{}^k = A_i B_j O_i D_j \sum_k \exp(-\beta c_{ij}{}^k) \tag{62}$$

Thus, equations (59)–(61) define a multi-mode distribution model. $T_{ij}{}^k$ and T_{ij} can be divided, using equations (59) and (62), to give the modal split as

$$\frac{T_{ij}{}^k}{T_{ij}} = \frac{\exp(-\beta c_{ij}{}^k)}{\sum_k \exp(-\beta c_{ij}{}^k)} \tag{63}$$

as the proportion travelling by mode k between i and j. Note that in the two-mode case, say with modes 1 and 2 representing public and private transport, a plot of, say, $T_{ij}{}^1/T_{ij}$ would give a curve which has the shape of the usual diversion curve.

Note further that the T_{ij} derived in (62) above can be wholly identified with the T_{ij} derived in the conventional gravity model in equation (36) provided that

$$\exp(-\beta c_{ij}) = \sum_k \exp(-\beta c_{ij}{}^k) \tag{64}$$

This equation is of the greatest importance because it shows how a composite measure of impedance, $\exp(-\beta c_{ij})$, or average generalized cost, c_{ij}, can be derived from the modal impedances $\exp(-\beta c_{ij}{}^k)$ where these are known individually. Such composite impedances are valuable in a variety of planning models, but past practice has been to use one of a number of arbitrary averaging procedures.

It should also be remarked that the modal split formula (63) is identical in form to that derived from a statistical approach to modal split using discriminant analysis (Quarmby, 1967). There could be complete identification if the generalized cost $c_{ij}{}^k$ could be identified with the discriminant function used by the statisticians. If such an identification can be made, then discriminant analysis would provide a method for determining the generalized costs.

Finally, let us examine the case where there is independent information on trip end estimation by mode. Suppose it is known that not only are there O_i trips in total from zone i, but that $O_i{}^k$ are by mode k, and that similarly there are $D_j{}^k$ trips by mode k into zone j. The constraints analogous to (53) and (54) can be written as

$$\sum_i T_{ij}{}^k = D_j{}^k \tag{65}$$

$$\sum_j T_{ij}{}^k = O_i{}^k \tag{66}$$

Sets of Lagrangian multipliers $\lambda_j^{(2)k}$, $\lambda_i^{(1)k}$ are now defined to be associated with these constraints. The maximization process is carried through in the usual way giving

$$T_{ij}{}^k = A_i{}^k B_j{}^k O_i{}^k D_j{}^k \exp(-\beta c_{ij}{}^k) \tag{67}$$

where

$$A_i^k = \left[\sum_j B_j^k D_j^k \exp(-\beta c_{ij}^k) \right]^{-1} \tag{68}$$

$$B_j^k = \left[\sum_i A_i^k O_i^k \exp(-\beta c_{ij}^k) \right]^{-1} \tag{69}$$

Notice that in this case, T_{ij}^k/T_{ij} is not as simple a ratio as with the previous one and no longer agrees with the results of the discriminant analysis approach.

An alternative result is obtained if the constraint (53) is used with (66) together with the usual cost constraint (55). (This represents the situation where trip generations are known by mode but trip attractions only in total for each zone.) This can be worked out in the usual way and the result is

$$T_{ij}^k = A_i^k B_j O_i^k D_j \exp(-\beta c_{ij}^k) \tag{70}$$

where

$$A_i^k = \left[\sum_j B_j D_j \exp(-\beta c_{ij}^k) \right]^{-1} \tag{71}$$

and

$$B_j = \left[\sum_i \sum_k A_i^k O_i^k \exp(-\beta c_{ij}^k) \right]^{-1} \tag{72}$$

The objective of obtaining these results and comparing them is to assess their importance for modal split applications in transportation studies, since in some cases it is assumed that trip ends by mode can be estimated, for example by regression analysis.

The model represented by equations (67)–(69), that is assuming that trip end modal split is known, is commonly used. Examination of the structure of these equations shows that they are completely separated for each mode: the trip ends are estimated separately, and then each mode is distributed separately. The weakness of this method arises in the need for forecasting as well as to explain the present, or in any situation where some of the parameters change: any aggregate switching of mode can only be brought about by changes in the O_i^k's and D_j^k's. By contrast, with the first multi-modal distribution model derived above, described by equations (53)–(55), the aggregate levels of modal choice are determined, given only total O_i's and D_j's by the relative costs, the c_{ij}^k's. This seems a much more fundamental mechanism than the separate mode methods, and is directly related to the sort of inter-modal comparison a traveller may be expected to make. Its connection with the discriminant analysis approach also adds weight to the view that it is a preferable method.

The intermediate method, that is the use of the model described by equations (70)–(72), falls between the two stools, and is probably not in current use anyway. However, it will be seen in the next section that a model of this structure will become important in the case where two types of traveller are considered: car owners and non-car owners.

3.4. *Extension of the multi-mode model to the case where some users have access to only a subset of all modes*

The multi-mode distribution models produced in Sections 3.3 assume implicitly that all travellers have access to all modes. There is at least one obvious case in real life where this is not so: non-car owners do not have the possibility of travel by car. This is important for current transportation study models, where usual practice is often to assume that trip ends for car owners, for example, can be separately estimated, and the trips separately distributed. This suffers from the same forecasting deficiency as the separate modal distributions of the previous section. It is, in fact, an exactly analogous assumption. However, once again the situation can be described by the appropriate constraints and a model can be produced which seems more appropriate to the situation. The basis of the constraint equations will be that, if any set of travellers should have only a subset of the modes available, then the trip end productions should be similarly categorized, but not trip attractions, so that all travellers compete for the same attractions, but non-car owners, for example, cannot generate trips by car. Thus trip productions would be generated separately for car owners and non-car owners, but only total trip attractions would be estimated.

The first step, however, is to get a general formulation of the problem and to develop an appropriate notation. Let n represent a class of travellers, and let $\gamma(n)$ be the set of modes available to travellers in category n. k will denote mode as usual, and

$$\sum_{k \in \gamma(n)}$$

denotes summation over the subset of modes k available to persons of the type n. The new constraints, embodying the principles discussed above, are then

$$\sum_{j} \sum_{k \in \gamma(n)} T_{ij}^{kn} = O_i^n \tag{73}$$

$$\sum_{i} \sum_{n} \sum_{k \in \gamma(n)} T_{ij}^{kn} = D_j \tag{74}$$

$$\sum_{i} \sum_{j} \sum_{n} \sum_{k \in \gamma(n)} T_{ij}^{kn} = C \tag{75}$$

where, in an obvious notation, T_{ij}^{kn} is the number of trips from i to j by mode k by traveller type n, O_i^n is the number of trip generations at i by travellers of type n, and other variables have been defined before. The maximand is now

$$\frac{T!}{\prod_{ijkn} T_{ij}^{kn}!} \tag{76}$$

subject to the constraints (73)–(75). Introduce Lagrangian multipliers $\lambda_i^{(1)n}$, $\lambda_j^{(2)}$ and β in the usual way and the maximizing condition is

$$-\log T_{ij}^{kn} - \lambda_i^{(1)n} - \lambda_j^{(2)} - \beta c_{ij}^k = 0 \tag{77}$$

so

$$T_{ij}^{kn} = \exp\left(-\lambda_i^{(1)n} - \lambda_j^{(2)} - \beta c_{ij}^k\right) \tag{78}$$

and writing

$$A_i^n = \exp\left(-\lambda_i^{(1)n}\right)/O_i^n \tag{79}$$

and

$$B_j = \exp\left(-\lambda_j^{(2)}\right)/D_j \tag{80}$$

the usual manipulation gives

$$T_{ij}^{kn} = A_i^n B_j O_i^n D_j \exp\left(-\beta c_{ij}^k\right) \tag{81}$$

where

$$A_i^n = \left[\sum_{j} \sum_{k \in \gamma(n)} B_j D_j \exp\left(-\beta c_{ij}^k\right)\right]^{-1} \tag{82}$$

and

$$B_j = \left[\sum_{j} \sum_{n} \sum_{k \in \gamma(n)} A_i^n O_i^n \exp\left(-\beta c_{ij}^k\right)\right]^{-1} \tag{83}$$

Note that we can now get the total inter-zonal trips by mode (by summing over n, denoted by $T_{ij}{}^k$), by traveller type (by summing over $k \in \gamma(n)$, denoted by $T_{ij}{}^n$), and in total (by summing over $k \in \gamma(n)$ and n, denoted by T_{ij}). Thus

$$T_{ij}{}^k = B_j \, D_j \left(\sum_n A_i{}^n O_i{}^n \right) \exp(-\beta c_{ij}{}^k) \tag{84}$$

$$T_{ij}{}^n = A_i{}^n B_j O_i{}^n D_j \sum_{k \in \gamma(n)} \exp(-\beta c_{ij}{}^k) \tag{85}$$

$$T_{ij} = B_j D_j \sum_n \sum_{k \in \gamma(n)} A_i{}^n O_i{}^n \exp(-\beta c_{ij}{}^k) \tag{86}$$

Note that in equation (85), the results for one person category n is linked to the other person type variables through the B_j's defined in (83). This arises because different person categories are competing for the same attractions.

A result which is suitable for a car owner/non-car owner split can now be obtained easily from these general equations.

3.5. *Derivation of the intervening opportunities model*

The intervening opportunities model was derived in the traditional way in the introduction to this paper and its main equation was derived as equation (15). It is of some interest to attempt to derive this using the new methodology, since, if this is possible, the gravity and opportunities models are related by this common base and can be compared in a new light.

Using the variables defined in the introduction, it is also possible to define in addition

$$S_{ij_\mu} = O_i U_{ij_\mu} \tag{87}$$

as the number of trips from i continuing beyond the μth ranked zone away from i. Note that, since

$$T_{ij_\mu} = S_{ij_{\mu-1}} - S_{ij_\mu} \tag{88}$$

the variables S_{ij_μ} define the new system as a possible alternative to T_{ij_μ}. To derive the opportunities model, the new method is applied to the variables S_{ij_μ}. Thus if S is the total numbers of states for a given distribution $\{S_{ij_\mu}\}$, then the maximand will be

$$\frac{S}{\prod_{ij_\mu} S_{ij_\mu}!}$$

It is now necessary to establish appropriate constraints. As seen earlier, the opportunities model does not have a constraint on trip attractions of the form of equation (4), but does need a constraint on trip generations of the form (3). For the variables S_{ij_μ}, the strictly analogous constraint to (3) is the inequality

$$S_{ij_\mu} \leqslant O_i \tag{89}$$

as there cannot be more trips continuing beyond a point from i than originally set out from i. If these are summed over j_μ to get a constraint of the form (3), the resulting equation is

$$\sum_{j_\mu} S_{ij_\mu} = k_i' O_i \tag{90}$$

where k_i' is some constant, and $1 \leqslant k_i' \leqslant N$, where N is the total number of zones. Finally, a constraint analogous to the gravity model cost constraint, equation (22), is needed.

The main assumption of the intervening opportunities model, as commonly stated, is that the number of trips between i and j is determined by the number of opportunities at j, and varies inversely as the number of intervening opportunities. This gives the clue for the cost constraint: to use intervening opportunities as a proxy for cost. Thus, if S_{ij_μ} trips are to be made beyond j_μ, then these will incur costs greater than those for trips which have been made to nearer zones. Suppose, then, that for trips from i, we take the number of opportunities passed as a measure of the cost of getting so far. Thus, the minimum cost for the remaining trips beyond j_μ is $A_{j_\mu} S_{ij_\mu}$. If this is summed over j_μ and then over all origin zones i, this gives a function which behaves in some ways like a total cost function, and the corresponding constraint is

$$\sum_i \sum_{j_\mu} A_{ij_\mu} S_{ij_\mu} = C \tag{91}$$

Since, as can be derived from the definitional equation (10),

$$A_{j_\mu} = \sum_{n=1}^{\mu} D_{j_n} \tag{92}$$

and, as can be derived from the definition of S_{ij_μ},

$$S_{ij_\mu} = \sum_{n=\mu+1}^{N} T_{ij_n} \tag{93}$$

where N is the total number of zones, it can easily be seen that the coefficient of T_{ij_μ} in the summation in equation (91) is [substituting for S_{ij_μ} from (93)]

$$(\mu-1) D_{j_1} + (\mu-2) D_{j_2} + \ldots + D_{j_{\mu-1}} \tag{94}$$

and so the opportunities passed contribute to the cost associated with a particular element of the trip matrix weighted by the number of times they have been "passed" or "have intervened".

Now, maximizing

$$\frac{S}{\prod_{ij_\mu} S_{ij_\mu}!}$$

subject to the constraints (90) and (91), introducing Lagragian multipliers $\lambda_i^{(1)}$ and L for these constraints, the most probable distribution occurs when

$$-\log S_{ij_\mu} - LA_{j_\mu} - \lambda_i^{(1)} = 0$$

so

$$S_{ij} = \exp(-LA_{j_\mu} - \lambda_i^{(1)}). \tag{95}$$

$\lambda_i^{(1)}$ can be obtained in the usual way by substituting from (95) into (90):

$$\exp(-\lambda_i^{(1)}) = \frac{k_i' O_i}{\sum_{j_\mu} \exp(-LA_{j_\mu})} \tag{96}$$

and so, writing

$$k_i = \frac{k_i'}{\sum_{j_\mu} \exp(-LA_{j_\mu})} \tag{97}$$

$$S_{ij_\mu} = k_i O_i \exp(-LAj_\mu) \tag{98}$$

268

and, using equation (88),

$$T_{ij_\mu} = k_i O_i [\exp(-LA_{j_{\mu-1}}) - \exp(-LA_{j_\mu})] \tag{99}$$

which is identical to equation (15). Thus, the main equation of the intervening opportunities model has been obtained using the new method.

This derivation has been made at the expense of using a rather strange cost constraint equation (92), and assuming a cost of getting from i to j_μ implied by equation (94). Perhaps this is an argument in itself for preferring the gravity model to the intervening opportunities model.

3.6. A distribution model of gravity type which uses intervening opportunities as a measure of cost

As a final example of the application of the new method, consider the distribution model which is obtained if intervening opportunities are used as a measure of cost, but not weighted in the form (94). This is perhaps a more plausible assumption.

Note that j_μ, as originally defined, is properly a function of i and should be written as $j_\mu(i)$. The assumption now proposed is

$$c_{ij} = A_{j_\mu(i)} \tag{100}$$

This cost can now be substituted in either of the two gravity models derived above, the so-called conventional model described by equations (36)–(38), or the so-called single competition term model described by equations (47)–(48) with $f(c_{ij})$ as $\exp(-\beta c_{ij})$. Thus, substituting for c_{ij} from (100), the double competition term "gravity/opportunity" model is

$$T_{ij_\mu} = a_i b_{j_\mu} O_i D_{j_\mu} \exp(-\beta A_{j_\mu(i)}) \tag{101}$$

$$a_i = \left[\sum_{j_\mu} b_{j_\mu} D_{j_\mu} \exp(-\beta A_{j_\mu(i)}) \right]^{-1} \tag{102}$$

$$b_{j_\mu} = \left[\sum_i a_i O_i \exp(-\beta A_{j_\mu(i)}) \right]^{-1} \tag{103}$$

Small a's and b's are used for the balancing factors to avoid confusion with the A_{j_μ}'s. The single competition term "gravity/opportunity" model is

$$T_{ij_\mu} = a_i O_i D_{j_\mu} \exp(-\beta A_{j_\mu(i)}) \tag{104}$$

$$a_i = \left[\sum_{j_\mu} D_{j_\mu} \exp(-\beta A_{j_\mu(i)}) \right]^{-1} \tag{105}$$

If it could be argued that (100) represents a better account of cost than (94), then the models represented by equations (101)–(103) and (104)–(105) may give better answers than the conventional intervening opportunities model. A test of these new models would be welcomed.

4. CONCLUSIONS

A new statistical theory of spatial distribution models has been demonstrated. This gives a new method for constructing such models to meet a wide variety of circumstances. It appears to show that the gravity model has a more plausible theoretical base than the intervening opportunities model. But above all, the new method offers a technique for extending conventional spatial distribution models to cover new situations which now often present serious problems, especially in transportation study models: for example, modal

A statistical theory of spatial distribution models   269

split, especially in the multi-mode case, and the necessity of separating car owners and non-car owners, but allowing them to compete for the same attractions. Finally, a comparison of the gravity models and intervening opportunities model leads to a suggestion of a completely different type of model for quite simple and conventional situations.

REFERENCES

Chicago Area Transportation Study (1960). Final report, Vol. II.
DIETER K. H. (1962). Distribution of work trips in Toronto. *Proc. Am. Soc. civ. Engrs* **88**, 9–28.
POPPER K. R. (1959). *The Logic of Scientific Discovery*. Hutchinson, London.
QUARMBY D. A. (1967). Choice of travel mode for the journey to work: some findings. *J. Transp. Economics and Policy*. To be published.
STOUFFLER S. A. (1940). Intervening opportunities: A theory relating mobility and distance. *Am. Soc. Rev.* **5**, No. 6.
TOLMAN R. C. (1938). *Principles of Statistical Mechanics*. Clarendon Press, Oxford.

Index